Customizing ASP.NET Core 6.0

Second Edition

Learn to turn the right screws to optimize ASP.NET Core applications for better performance

Jürgen Gutsch

BIRMINGHAM—MUMBAI

Customizing ASP.NET Core 6.0
Second Edition

Associate Group Product Manager: Pavan Ramchandani
Senior Editor: Mark Dsouza
Content Development Editor: Divya Vijayan
Technical Editor: Saurabh Kadave
Copy Editor: Safis Editing
Project Coordinator: Rashika Ba
Proofreader: Safis Editing
Indexer: Sejal Dsilva
Production Designer: Prashant Ghare
Marketing Coordinator: Anamika Singh

First published: January 2021

Second edition: December 2021

Production reference: 2270122

Published by Packt Publishing Ltd.
Livery Place
35 Livery Street
Birmingham
B3 2PB, UK.

ISBN 978-1-80323-360-4

www.packt.com

*To my family, friends, colleagues, and the .NET developer community,
who supported and encouraged me to do this work. To my three sons,
who gave me the power to complete it.*

– Jürgen Gutsch

Foreword

I have known Jürgen for more than 7 years and he never ceases to impress me with his knowledge and his experience in software development and web development, his expertise in ASP.NET Core and Azure solutions, and his consultancy and software architecture skills. Jürgen has been involved with ASP.NET since the first releases and has a vast amount of experience with all the .NET and .NET Core versions. His journey has now moved on to Azure and application development in the cloud using the latest ASP. NET Core technologies, Azure DevOps, and cloud-based solutions.

Jürgen is one of the leading community experts in the Swiss and German region and is extremely active in this field. You will always find him present at local conferences, .NET Core meetups, and Microsoft events, or organizing these events himself. He has been a Microsoft MVP for Developer Technologies since 2015 and I have had the pleasure to develop and learn from him by implementing solutions using some of the early versions and pre-releases of ASP.NET Core or solutions such as ASP.NET Core health checks. He is always helping the .NET community by giving them a chance to speak at meetups and offering speakers a platform. I did my first .NET user group talk in Basel with his help and support.

Above all, he is always there as a friend and I can always ask for support or his help in solving problems, and he helps make the .NET community in Switzerland so strong, helpful, and easy for others to join and improve.

Damien Bowden

Blogger on damienbod.com, Microsoft MVP for Developer Technologies, and Senior Software Consultant.

Well known on Twitter as @damien_bod

Personally, I believe we should always try to improve. I often ask myself three questions after I do anything:

- How could I have done that faster?
- How could I have done that better?
- How do I make sure I don't have to do that again?

By asking these questions, we can always improve in everything we do. That last question involves strategic scale, which is trickier (you usually have to improve with #1 and #2 to get to an answer for #3). For example, if you clean a coffee machine every night, you might ask yourself, "How can I clean that faster, how can I make it even cleaner, and how do I make sure I don't have to do that again?" You might not be able to answer that third question for quite a while, but pretty soon, you'd be an expert, in speed and quality, at cleaning coffee pots (possibly the best in the world). As you work to answer the third question, you'd start by training the people around you, which means more people would be cleaning (and you'd be doing it less). Finally, you'd likely get promoted to manager (since you're already training everyone). Thus, you would no longer have to clean coffee pots!

"Sounds good, Ed, but what does this have to do with code?" Great question, me! To answer that, let's head to Grace Hopper, one of our key leaders in pioneering computer technology...

> *"The most dangerous phrase in the language is, 'We've always done it this way.'"*
>
> *– Grace Hopper*

As Grace implies, our greatest danger is when we stop learning and ultimately stop improving our products and our code! As the CEO of Microsoft eloquently said:

> *"Always keep learning. You stop doing useful things if you don't learn."*
>
> *– Satya Nadella*

And if we don't learn and improve our code, then our product remains stagnant. But if we keep learning and improving, then not only does our product get faster (which answers question #1) and more useful (question #2), but it also makes us far more valuable to our product and company! You might want to keep coding for the rest of your life, but if you're providing that much value to your team, then there will be many opportunities to train them and potentially even manage them (which means coding less often, if that's appealing to you). At the very least, it will open many opportunities in your career!

In *Customizing ASP.NET Core 6.0*, Microsoft MVP Jürgen Gutsch brings his nearly 20 years of experience as a .NET and ASP.NET web developer to this book. Following Grace Hopper's and Satya Nadella's pleas for us to never stop learning and improving, Jürgen shows you how to do that to customize, improve, and optimize your ASP.NET Core applications. He shows you how to customize various aspects of your application, including the hosting layer (logging, app configuration, dependency injection, HTTPS configuration with Kestrel, and hosting models), the middleware layer, the routing layer (endpoint routing), the MVC layer (customizing and configuring identity management), and the Web API layer (content negotiation and managing inputs)! By the time you're done with this book, you'll have a richer understanding of .NET development and how you can get the most out of your applications.

You'll find that each chapter begins with a high-level architecture diagram that visually explains the chapter's architecture and which layer of the architecture the chapter is referring to. Use these diagrams to create a strong mental picture of how the architectural layers build on each other. Even if you understand the topic of customization in each chapter, you should try out the code to become more familiar with the various opportunities that you'll have to improve your application. Whether you're building a custom logger, customizing dependency injection, or enabling customers to manage their profiles, this book thoroughly goes through each step of app customization and optimization. As Satya Nadella implores, here's your chance to "always keep learning." Jump in and join Jürgen Gutsch on this learning journey.

Ed Price

Senior Program Manager of Architectural Publishing at Microsoft

Azure Architecture Center (`http://aka.ms/Architecture`) Co-author of six books, including *The Azure Cloud Native Architecture Mapbook* and *ASP.NET Core 5 for Beginners* (both from *Packt*)

Contributors

About the author

Jürgen Gutsch is a .NET-addicted web developer. He has worked with .NET and ASP.NET since the early versions in 2002. Before that, he wrote server-side web applications using classic ASP. He is also an active part of the .NET developer community. Jürgen writes for the dotnetpro magazine, one of the most popular German-speaking developer magazines. He also publishes articles in English on his blog, ASP.NET Hacker, and contributes to several open source projects. Jürgen has been a Microsoft MVP since 2015.

The best way to contact him is by using Twitter: @sharpcms.

He works as a developer, consultant, and trainer for the digital agency YOO Inc., located in Basel, Switzerland. YOO Inc. serves national as well as international clients and specializes in creating custom digital solutions for distinct business needs.

I want to thank my boss, my colleagues, and my employer for their support and motivation.

About the reviewer

Toi B. Wright has been obsessed with ASP.NET for almost 20 years. She is the founder and president of the Dallas ASP.NET User Group. She has been a Microsoft MVP in ASP.NET for 16 years and is also an ASPInsider. She is an experienced full-stack software developer, book author, courseware author, speaker, and community leader with over 25 years of experience. She has a B.S. in computer science and engineering from the Massachusetts Institute of Technology (MIT) and an MBA from Carnegie Mellon University (CMU).

You can find her on Twitter: @misstoi.

Table of Contents

8

Writing Custom Middleware

9

Working with Endpoint Routing

10

Customizing ASP.NET Core Identity

11

Configuring Identity Management

12
Content Negotiation Using a Custom OutputFormatter

13
Managing Inputs with Custom ModelBinder

14
Creating a Custom ActionFilter

15
Working with Caches

16

Creating Custom TagHelper

Preface

ASP.NET Core is the most powerful web framework provided by Microsoft and is full of hidden features that make it even more powerful and useful.

Your application should not be made to match the framework; your framework should be able to do what your application really needs. With this book, you will learn how to find the hidden screws you can turn to get the most out of the framework.

Developers working with ASP.NET Core will be able to put their knowledge to work with this practical guide to customizing ASP.NET Core. The book provides a hands-on approach to implementation and its associated methodologies that will have you up and running and productive in no time.

This book is a compact collection of default ASP.NET Core behaviors you might want to change and step-by-step explanations of how to do so.

By the end of this book, you will know how to customize ASP.NET Core to get an optimized application out of it according to your individual needs.

ASP.NET Core architecture overview

To follow the next chapters, you should be familiar with the base architecture of ASP.NET Core and its components. This book tackles almost all of the components of the architecture.

The following figure shows the base architecture overview of ASP.NET Core 6.0. Let's quickly go through the components shown here from the bottom to the top layer:

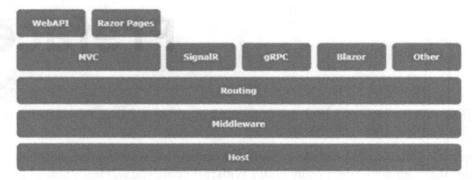

At the bottom, there is the **Host** layer. This is the layer that bootstraps the web server and all the stuff that is needed to start up an ASP.NET Core application, including logging, configuration, and the service provider. This layer creates the actual request objects and their dependencies that are used in the layers above.

The next layer above **Host** is the **Middleware** layer. This layer works with the request object or manipulates it. This attaches the middleware to the request object. It executes the middleware for things such as error handling, authenticating HSTS, CORS, and so on.

Above that, there is the **Routing** layer, which routes the request to the endpoints depending on the route patterns defined. Endpoint routing is the new player from ASP.NET Core 3.1 and separates routing from the UI layers above to enable routing for different endpoints, including Blazor, gRPC, and SignalR. As a reminder: in previous versions of ASP.NET Core, routing was part of the MVC layer, and every other UI layer needed to implement its own routing.

The actual endpoints are provided by the fourth layer, the UI layer, which contains the well-known UI frameworks **Blazor**, **gRPC**, **SignalR**, and **MVC**. This is where you will do most of your work as an ASP.NET Core developer.

Lastly, above **MVC**, you will find **WebAPI** and **Razor Pages**.

What is covered in this book?

This book doesn't cover all the topics mentioned in the architecture overview. This book covers most topics of the host layer because that is the layer that contains the most things you might need to customize. This book tackles middleware and routing, as well as MVC features and some more WebAPI topics, where you can do some magic tricks.

At the beginning of each chapter, we will indicate to which level the topic belongs.

What is not covered and why?

This book doesn't cover Razor Pages, SignalR, gRPC, and Blazor.

The reason is that gRPC and SignalR are already very specialized, and don't really need to be customized. Blazor is a new member of the ASP.NET Core family and is not widely used yet. Also, the author isn't familiar enough with Blazor to know all the screws to customize it. Razor Pages is on top of the MVC framework and customizations for MVC are also valid for Razor Pages.

Who this book is for

This book is for web developers working with ASP.NET Core, who might need to change default behaviors to get things done. Readers should have basic knowledge of ASP. NET Core and C#, since this book does not cover the basics of those technologies. Readers should also have a good knowledge of Visual Studio, Visual Studio Code, or any other code editor that supports ASP.NET Core and C#.

What this book covers

Chapter 1, *Customizing Logging*, teaches you how to customize the logging behavior and how to add a custom logging provider.

Chapter 2, *Customizing App Configuration*, helps you understand how to use different configuration sources and add custom configuration providers.

Chapter 3, *Customizing Dependency Injection*, teaches you how **Dependency Injection (DI)** works and how to use a different DI container.

Chapter 4, *Configuring and Customizing HTTPS with Kestrel*, looks into configuring HTTPS differently.

Chapter 5, *Configuring WebHostBuilder*, helps you understand how to set up configuration on the hosting layer.

Chapter 6, *Using Different Hosting Models*, teaches you about different types of hosting on different platforms.

Chapter 7, *Using IHostedService and BackgroundService*, makes you understand how to execute tasks in the background.

Chapter 8, *Writing Custom Middleware*, deals with the HTTP context using middleware.

Chapter 9, Working with Endpoint Routing, helps you understand how to use the new routing to provide custom endpoints.

Chapter 10, Customizing ASP.NET Core Identity, explains how to extend the application's user properties and helps you to change the Identity UI.

Chapter 11, Configuring Identity Management, helps you to manage your users and their roles.

Chapter 12, Content Negotiation Using a Custom OutputFormatter, teaches you how to output different content types based on the HTTP Accept header.

Chapter 13, Managing Inputs with Custom ModelBinder, helps you create input models with different types of content.

Chapter 14, Creating a Custom ActionFilter, covers aspect-oriented programming using ActionFilter.

Chapter 15, Working with Caches, helps you to make your application faster.

Chapter 16, Creating Custom TagHelper, enables you to simplify the UI layer by creating TagHelper.

To get the most out of this book

Readers should have basic knowledge of ASP.NET Core and C#, as well as Visual Studio, Visual Studio Code, or any other code editor that supports ASP.NET Core and C#.

Software/hardware covered in the book	Operating system requirements
ASP.NET 6.0	Windows, macOS, or any Linux distro

You should install the latest .NET 6.0 SDK on your machine. Please find the latest version at `https://dotnet.microsoft.com/download/dotnet-core/`.

Feel free to use any code editor you like that supports ASP.NET Core and C#. We recommend using Visual Studio Code (`https://code.visualstudio.com/`), which is available on all platforms and is used by the author of this book.

All the projects in this book will be created using a console, Command Prompt, shell, or PowerShell. Feel free to use whatever console you are comfortable with. The author uses Windows Command Prompt, hosted in the cmder shell (`https://cmder.net/`). We don't recommend using Visual Studio to create the projects, because the basic configuration might change, and the web projects will start on a different port than described in this book.

Are you stuck with .NET Core 3.1, or .NET 5.0? If you are not able to use .NET 6.0 on your machine for whatever reason, all the samples are also available and work with .NET Core 3.1 and .NET 5.0. Some chapters contain comparisons to .NET 5.0 whenever there are differences to .NET 6.0.

If you are using the digital version of this book, we advise you to type the code yourself or access the code from the book's GitHub repository (a link is available in the next section). Doing so will help you avoid any potential errors related to the copying and pasting of code.

Download the example code files

You can download the example code files for this book from GitHub at `https://github.com/PacktPublishing/Customizing-ASP.NET-Core-6.0-Second-Edition`. If there's an update to the code, it will be updated in the GitHub repository.

We also have other code bundles from our rich catalog of books and videos available at `https://github.com/PacktPublishing/`. Check them out!

Download the color images

We also provide a PDF file that has color images of the screenshots and diagrams used in this book. You can download it here: `https://static.packt-cdn.com/downloads/9781803233604_ColorImages.pdf`.

Conventions used

There are a number of text conventions used throughout this book.

`Code in text`: Indicates code words in text, database table names, folder names, filenames, file extensions, pathnames, dummy URLs, user input, and Twitter handles. Here is an example: "You can use `ConfigureAppConfiguration` to configure the app configuration."

A block of code is set as follows:

```
builder.Configuration.AddJsonFile(
        "appsettings.json",
        optional: false,
        reloadOnChange: true);
```

When we wish to draw your attention to a particular part of a code block, the relevant lines or items are set in bold:

```
builder.Logging.AddConfiguration(builder.Configuration.
GetSection("Logging"));
builder.Logging.AddConsole();
builder.Logging.AddDebug();
```

Any command-line input or output is written as follows:

```
cd LoggingSample
code .
```

Bold: Indicates a new term, an important word, or words that you see onscreen. For instance, words in menus or dialog boxes appear in **bold**. Here is an example: "Click on **Register** in the upper left-hand corner, you will see the following page."

> **Tips or important notes**
> Appear like this.

Get in touch

Feedback from our readers is always welcome.

General feedback: If you have questions about any aspect of this book, email us at customercare@packtpub.com and mention the book title in the subject of your message.

Errata: Although we have taken every care to ensure the accuracy of our content, mistakes do happen. If you have found a mistake in this book, we would be grateful if you would report this to us. Please visit www.packtpub.com/support/errata and fill in the form.

Piracy: If you come across any illegal copies of our works in any form on the internet, we would be grateful if you would provide us with the location address or website name. Please contact us at copyright@packt.com with a link to the material.

If you are interested in becoming an author: If there is a topic that you have expertise in and you are interested in either writing or contributing to a book, please visit authors.packtpub.com.

Share Your Thoughts

Once you've read *Customizing ASP.NET Core 6.0*, we'd love to hear your thoughts! Scan the QR code below to go straight to the Amazon review page for this book and share your feedback.

https://packt.link/r/1803233605

Your review is important to us and the tech community and will help us make sure we're delivering excellent quality content.

1

Customizing Logging

In this chapter, the first in this book about customizing **ASP.NET Core**, you will see how to customize **logging**. The default logging only writes to the console or the debug window. This is quite good for the majority of cases, but sometimes you need to log to a sink, such as a file or a database. Or, perhaps you want to extend the logger with additional information. In these cases, you need to know how to change the default logging.

In this chapter, we will be covering the following topics:

- Configuring logging
- Creating a custom logger
- Plugging in an existing third-party logger provider

The topics in this chapter refer to the hosting layer of the ASP.NET Core architecture:

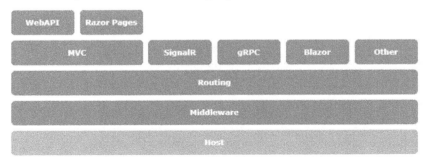

Figure 1.1 – The ASP.NET Core architecture

Technical requirements

To follow the descriptions in this chapter, you will need to create an ASP.NET Core MVC application. To do this, open your console, shell, or Bash terminal, and change to your working directory. Then, use the following command to create a new MVC application:

```
dotnet new mvc -n LoggingSample -o LoggingSample
```

Now, open the project in **Microsoft Visual Studio** by double-clicking the project file, or in Visual Studio Code, by typing the following command in the already-open console:

```
cd LoggingSample
code .
```

All of the code samples in this chapter can be found in the **GitHub** repository for this book at https://github.com/PacktPublishing/Customizing-ASP. NET-Core-6.0-Second-Edition/tree/main/Chapter01.

Configuring logging

In previous versions of ASP.NET Core (that is, before version 2.0), logging was configured in Startup.cs. As a reminder, since version 2.0, the Startup.cs file has been simplified, and a lot of configurations have been moved to the default WebHostBuilder, which is called in Program.cs. Also, logging was moved to the default WebHostBuilder.

In ASP.NET Core 3.1 and later versions, the Program.cs file gets more generic, and IHostBuilder will be created first. IHostBuilder is pretty useful for bootstrapping an application without all of the ASP.NET web stuff. We'll learn a lot more about IHostBuilder later on in this book. With this IHostBuilder, we create IWebHostBuilder to configure ASP.NET Core. In ASP.NET Core 3.1 and later versions, we get IWebHostBuilder with the webBuilder variable:

```
public class Program
{
    public static void Main(string[] args)
    {
        CreateHostBuilder(args).Build().Run();
    }

    public static IHostBuilder CreateHostBuilder(
```

```
  string[]args) =>
    Host.CreateDefaultBuilder(args)
        .ConfigureWebHostDefaults(webBuilder =>
        {
            webBuilder.UseStartup<Startup>();
        });
}
```

In ASP.NET Core 6.0, Microsoft introduced the **minimal API** approach that simplifies the configuration a lot. This approach doesn't use the Startup file and adds all of the configurations to the Program.cs file instead. Let's see what this looks like:

```
var builder = WebApplication.CreateBuilder(args);

// Add services to the container.
builder.Services.AddControllersWithViews();

var app = builder.Build();

// The rest of the file isn't relevant for this chapter
```

In ASP.NET Core, you are able to override and customize almost everything. This includes logging. IWebHostBuilder has a lot of extension methods that allow us to override the default behavior of different features. To override the default settings for logging, we need to use the ConfigureLogging method. The following code snippet shows almost exactly the same logging as was configured inside the ConfigureWebHostDefaults() method:

```
Host.CreateDefaultBuilder(args)
    .ConfigureWebHostDefaults(webBuilder =>
    {
        webBuilder
            .ConfigureLogging((hostingContext, logging) =>
            {
                logging.AddConfiguration(
                    hostingContext.Configuration.GetSection(
                        "Logging"));
                logging.AddConsole();
                logging.AddDebug();
```

```
            })
            .UseStartup<Startup>();
```

Using the minimal API approach, we don't need the `ConfigureLogging` method anymore, and we can use the `Logging` property of `WebApplicationBuilder` directly:

```
builder.Logging.AddConfiguration(builder.Configuration.
GetSection("Logging"));
builder.Logging.AddConsole();
builder.Logging.AddDebug();
```

Now that we've seen how to configure logging, let's look at building a custom logger.

Creating a custom logger

To demonstrate a custom logger, let's use a small, simple logger I created that is able to colorize log entries with a specific log level in the console. This logger is called `ColoredConsoleLogger`, and it will be created and added using `LoggerProvider`, which we also need to write for ourselves. To specify the color and the log level to colorize, we need to add a configuration class.

In the next snippets, all three parts (`Logger`, `LoggerProvider`, and `Configuration`) are shown:

1. Let's create the configuration class of our logger in a new file called `CustomLogger.cs` in the same folder as the `Program.cs` file. Add the following using statement at the top of the file:

   ```
   namespace LoggingSample;
   ```

   We will call it `ColoredConsoleLoggerConfiguration`. This class contains three properties to define – `LogLevel`, `EventId`, and `Color` - that can be set:

   ```
   public class ColoredConsoleLoggerConfiguraticn
   {
       public LogLevel LogLevel { get; set; } =
           LogLevel.Warning;
       public int EventId { get; set; } = 0;
       public ConsoleColor Color { get; set; } =
           ConsoleColor.Yellow;
   }
   ```

2. Next, we need a provider to retrieve the configuration and create the actual logger instance:

```
public class ColoredConsoleLoggerProvider :
ILoggerProvider
{
    private readonly ColoredConsoleLoggerConfiguration
      _config;
    private readonly ConcurrentDictionary<string,
      ColoredConsoleLogger> _loggers =
        new ConcurrentDictionary<string,
          ColoredConsoleLogger>();

    public ColoredConsoleLoggerProvider
      (ColoredConsoleLoggerConfiguration config)
    {
        _config = config;
    }

    public ILogger CreateLogger(string categoryName)
    {
        return _loggers.GetOrAdd(categoryName, name =>
          new ColoredConsoleLogger(name, _config));
    }

    public void Dispose()
    {
        _loggers.Clear();
    }
}
```

Don't forget to add a using statement for System.Collections. Concurrent.

3. The third class is the actual logger we want to use:

```
public class ColoredConsoleLogger : ILogger
{
    private static object _lock = new Object();
```

```csharp
    private readonly string _name;
    private readonly ColoredConsoleLoggerConfiguration
      _config;

    public ColoredConsoleLogger(
        string name,
        ColoredConsoleLoggerConfiguration config)
    {
        _name = name;
        _config = config;
    }

    public IDisposable BeginScope<TState>(TState
      state)
    {
        return null;
    }

    public bool IsEnabled(LogLevel logLevel)
    {
        return logLevel == _config.LogLevel;
    }

    public void Log<TState>(
        LogLevel logLevel,
        EventId eventId,
        TState state,
        Exception exception,
        Func<TState, Exception, string> formatter)
    {
        if (!IsEnabled(logLevel))
        {
            return;
        }

        lock (_lock)
```

```
        {
            if (_config.EventId == 0 ||
                _config.EventId == eventId.Id)
            {
                var color = Console.ForegroundColor;
                Console.ForegroundColor =
                  _config.Color;
                Console.Write($"{logLevel} - ");
                Console.Write($"{eventId.Id} - {_name}
                            - ");
                Console.Write($"{formatter(state,
                            exception)}\n");
                Console.ForegroundColor = color;
            }
        }
    }
}
```

We now need to lock the actual console output – this is because we will encounter some race conditions where incorrect log entries get colored with the wrong color, as the console itself is not really thread-safe.

4. After this is done, we can start to plug in the new logger to the configuration in Program.cs:

```
builder.Logging.ClearProviders();
var config = new ColoredConsoleLoggerConfiguration
{
    LogLevel = LogLevel.Information,
    Color = ConsoleColor.Red
};
builder.Logging.AddProvider(new
    ColoredConsoleLoggerProvider(config));
```

You might need to add a using statement to the LoggerSample namespace.

If you don't want to use the existing loggers, you can clear all the logger providers added previously. Then, we call AddProvider to add a new instance of our ColoredConsoleLoggerProvider class with the specific settings. We could also add some more instances of the provider with different settings.

This shows how you could handle the log levels in a different way. You could use this approach to send emails regarding hard errors or to log debug messages to a different log sink from regular informational messages, and much more.

Figure 1.2 shows the colored output of the previously created custom logger:

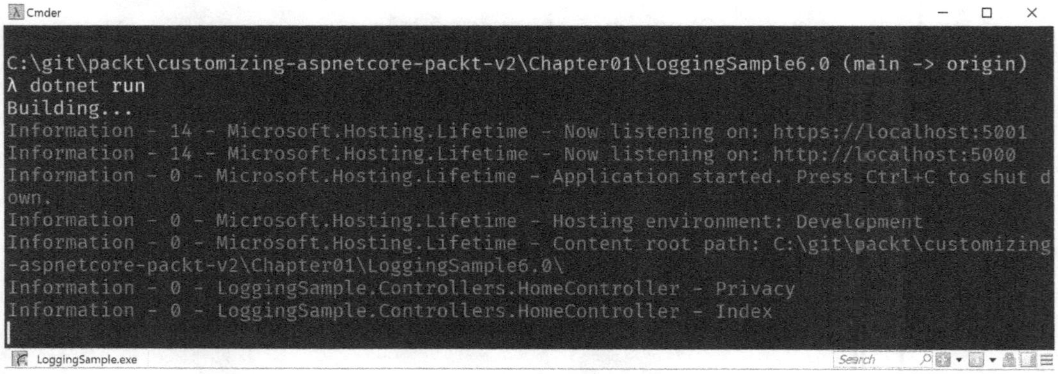

Figure 1.2 – A screenshot of the custom logger

In many cases, it doesn't make sense to write a custom logger, as many good third-party loggers are already available, such as ELMAH, log4net, and NLog. In the next section, we will see how to use NLog in ASP.NET Core.

Plugging in an existing third-party logger provider

NLog was one of the very first available as a **.NET Standard** library, and it can be used in ASP.NET Core. NLog also already provides a logger provider to easily plug into ASP.NET Core.

You will find NLog via **NuGet** (https://www.nuget.org/packages/NLog.Web.AspNetCore) and on GitHub (https://github.com/NLog/NLog.Web). Even if NLog is not yet explicitly available for ASP.NET Core 6.0, it will still work with version 6.0:

1. We need to add an NLog.Config file that defines two different sinks to log all standard messages in a single log file and custom messages only in another file. Since this file is too long to print, you can view it or download it directly from GitHub: https://github.com/PacktPublishing/Customizing-ASP.NET-Core-6.0-Second-Edition/blob/main/Chapter01/LoggingSample6.0/NLog.Config

2. We then need to add the NLog ASP.NET Core package from NuGet:

```
dotnet add package NLog.Web.AspNetCore
```

> **Important Note**
> Be sure that you are in the project directory before you execute the preceding command!

3. Now, you only need to clear all the other providers in the ConfigureLogging method in Program.cs and to use NLog with IWebHostBuilder using the UseNLog() method:

```
Host.CreateDefaultBuilder(args)
    .ConfigureWebHostDefaults(webBuilder =>
    {
        webBuilder
            .ConfigureLogging((hostingContext,
              logging) =>
            {
                logging.ClearProviders();
                logging.SetMinimumLevel(
                  LogLevel.Trace);
            })
            .UseNLog()
            .UseStartup<Startup>();
    });
```

Using the minimal API, it looks much simpler:

```
using NLog.Web;

var builder = WebApplication.CreateBuilder(args);

builder.Logging.ClearProviders();
builder.Logging.SetMinimumLevel(LogLevel.Trace);
builder.WebHost.UseNLog();
```

Here, you can add as many logger providers as you require.

That covers using an existing third-party logger. Let's now recap what we've covered in this chapter.

Summary

The good thing about hiding the basic configuration of an application is that it allows you to clean up the newly scaffolded projects and to keep the actual start as simple as possible. The developer is able to focus on the actual features. However, the more the application grows, the more important logging becomes. The default logging configuration is easy and it works like a charm, but in production, you need a persisted log to see errors from the past. Therefore, you need to add a custom logging configuration or a more flexible third-party logger, such as NLog or log4net.

You will learn more about how to configure ASP.NET Core 6.0 in the next chapter.

2
Customizing App Configuration

This second chapter is about application configuration, how to use it, and how to customize the ASP.NET configuration to employ different ways to configure your app. Perhaps you already have an existing **Extensible Markup Language (XML)** configuration or want to share a **YAML Ain't Markup Language (YAML)** configuration file over different kinds of applications. Sometimes, it also makes sense to read configuration values out of a database.

In this chapter, we will be covering the following topics:

- Configuring the configuration
- Using typed configurations
- Configuration using **Initialization (INI)** files
- Configuration providers

The topics in this chapter refer to the hosting layer of the ASP.NET Core architecture:

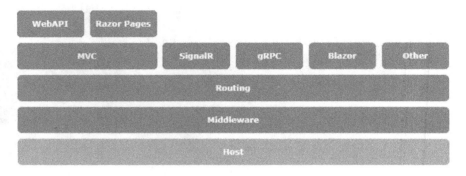

Figure 2.1 – ASP.NET Core architecture

Technical requirements

To follow the descriptions in this chapter, you will need to create an ASP.NET Core **Model-View-Controller** (**MVC**) application. Open your console, shell, or Bash terminal, and change to your working directory. Use the following command to create a new MVC application:

```
dotnet new mvc -n ConfigureSample -o ConfigureSample
```

Now, open the project in Visual Studio by double-clicking the project file or, in **Visual Studio Code** (**VS Code**), by typing the following command in the already open console:

```
cd ConfigureSample
code .
```

All of the code samples in this chapter can be found in the GitHub repository for this book at https://github.com/PacktPublishing/Customizing-ASP.NET-Core-6.0-Second-Edition/tree/main/Chapter02.

Configuring the configuration

Let's start by looking at how to configure your various configuration options.

Since ASP.NET Core 2.0, the configuration is hidden in the default configuration of WebHostBuilder and is no longer part of Startup.cs. This helps to keep the startup clean and simple.

In ASP.NET Core 3.1 up to ASP.NET Core 5.0, the code looks like this:

```
// ASP.NET Core 3.0 and later
public class Program
{
    public static void Main(string[] args)
    {
        CreateWebHostBuilder(args).Build().Run();
    }
    public static IHostBuilder CreateHostBuilder(string[]
      args) =>
        Host.CreateDefaultBuilder(args)
            .ConfigureWebHostDefaults(webBuilder =>
            {
                webBuilder.UseStartup<Startup>();
            }
}
```

In ASP.NET Core 6.0, Microsoft introduced the minimal **application programming interface (API)** approach that simplifies the configuration a lot. This doesn't use `Startup` and adds all the configuration in the `Program.cs` file. Let's see how it looks here:

```
Var builder = WebApplication.CreateBuilder(args);

// Add services to the container.
builder.Services.AddControllersWithViews();

var app = builder.Build();

// The rest of the file isn't relevant for this chapter
```

Fortunately, in both versions, you are also able to override the default settings to customize the configuration in the way you need it. In both versions, we extend `IWebHostBuilder` with the `ConfigureAppConfiguration()` method where the magic will happen.

This is what the configuration looks like in ASP.NET Core 3.1 and ASP.NET Core 5.0:

```
Host.CreateDefaultBuilder(args)
    .ConfigureWebHostDefaults(webBuilder =>
    {
        webBuilder
          .ConfigureAppConfiguration((builderContext,
            config) =>
        {
            // configure configuration here
        })
        .UseStartup<Startup>();
    });
```

This is what the code looks like when using the minimal API approach. You also can use `ConfigureAppConfiguration` to configure the app configuration:

```
builder.WebHost.ConfigureAppConfiguration((builderContext,
config) =>
{
    // configure configuration here
});
```

But there is a much simpler approach, by accessing the `Configuration` property of the builder:

```
builder.Configuration.AddJsonFile(
    "appsettings.json",
    optional: false,
    reloadOnChange: true);
```

When you create a new ASP.NET Core project, you will already have `appsettings.json` and `appsettings.Development.json` configured. You can, and should, use these configuration files to configure your app; this is the preconfigured way, and most ASP.NET Core developers will look for an `appsettings.json` file to configure the application. This is absolutely fine and works pretty well.

The following code snippet shows the encapsulated default configuration to read the appsettings.json files:

```
var env = builder.Environment;
builder.Configuration.SetBasePath(env.ContentRootPath);
builder.Configuration.AddJsonFile(
    "appsettings.json",
    optional: false,
    reloadOnChange: true);
builder.Configuration.AddJsonFile(
    $"appsettings.{env.EnvironmentName}.json",
    optional: true,
    reloadOnChange: true);
builder.Configuration.AddEnvironmentVariables();
```

This configuration also sets the base path of the application and adds the configuration via environment variables.

Whenever you customize the application configuration, you should add the configuration via environment variables as a final step, using the AddEnvironmentVariables() method. The order of the configuration matters and the configuration providers that you add later on will override the configurations added previously. Be sure that the environment variables always override the configurations that are set via a file. This way, you also ensure that the configuration of your application on an Azure App Service will be passed to the application as environment variables.

IConfigurationBuilder has a lot of extension methods to add more configurations, such as XML or INI configuration files and in-memory configurations. You can find additional configuration providers built by the community to read in YAML files, database values, and a lot more. In an upcoming section, we will see how to read INI files. First, we will look at using typed configurations.

Using typed configurations

Before trying to read INI files, it makes sense for you to see how to use typed configurations instead of reading the configuration via IConfiguration, key by key.

To read a typed configuration, you need to define the type to configure. I usually create a class called `AppSettings`, as follows:

```
namespace ConfigureSample;

public class AppSettings
{
    public int Foo { get; set; }
    public string Bar { get; set; }
}
```

This is a simple **Plain Old CLR Object (POCO)** class that will only contain the application setting values, as illustrated in the following code snippet. These classes can then be filled with specific configuration sections inside the `ConfigureServices` method in `Startup.cs` until ASP.NET Core 5.0:

```
services.Configure<AppSettings>
    (Configuration.GetSection("AppSettings"));
```

Using the minimal API approach, you need to configure the `AppSettings` class, like this:

```
builder.Services.Configure<AppSettings>(
    builder.Configuration.GetSection("AppSettings"));
```

This way, the typed configuration also gets registered as a service in the **dependency injection (DI)** container and can be used everywhere in the application. You are able to create different configuration types for each configuration section. In most cases, one section should be fine, but sometimes it makes sense to divide the settings into different sections. The next snippet shows how to use the configuration in an MVC controller:

```
using Microsoft.Extensions.Options;
// ...
public class HomeController : Controller
{
    private readonly AppSettings _options;

    public HomeController(IOptions<AppSettings> options)
    {
        _options = options.Value;
```

```
    }

    public IActionResult Index()
    {
        ViewData["Message"] = _options.Bar;
        return View();
    }
```

IOptions<AppSettings> is a wrapper around our AppSettings type, and the Value property contains the actual instance of AppSettings, including the values from the configuration file.

To try reading the settings in, the appsettings.json file needs to have the AppSettings section configured, otherwise the values are null or not set. Let's now add the section to the appsettings.json file, as follows:

```
{
    "Logging": {
        "LogLevel": {
            "Default": "Warning"
        }
    },
    "AllowedHosts": "*",
    "AppSettings": {
        "Foo": 123,
        "Bar": "Bar"
    }
}
```

Next, we'll examine how INI files can be used for configuration.

Configuration using INI files

To also use INI files to configure the application, you will need to add the INI configuration inside the ConfigureAppConfiguration() method in Program.cs, as follows:

```
builder.Configuration.AddIniFile(
    "appsettings.ini",
```

```
    optional: false,
    reloadOnChange: true);
builder.Configuration.AddJsonFile(
    $"appsettings.{env.EnvironmentName}.ini",
    optional: true,
    reloadOnChange: true);
```

This code loads the INI files the same way as the **JavaScript Object Notation** (**JSON**) configuration files. The first line is a required configuration, and the second line is an optional configuration depending on the current runtime environment.

The INI file could look like this:

```
[AppSettings]
Bar="FooBar"
```

As you can see, this file contains a section called AppSettings and a property called Bar.

Earlier, we said that the order of the configuration matters. If you add the two lines to configure via INI files after the configuration via JSON files, the INI files will override the settings from the JSON files. The Bar property gets overridden with "FcoBar" and the Foo property stays the same because it will not be overridden. Also, the values out of the INI file will be available via the typed configuration created previously.

Every other configuration provider will work the same way. Let's now see how a configuration provider will look.

Configuration providers

A configuration provider is an implementation of IConfigurationProvider that is created by a configuration source, which is an implementation of IConfigurationSource. The configuration provider then reads the data from somewhere and provides it via Dictionary.

To add a custom or third-party configuration provider to ASP.NET Core, you will need to call the Add method on ConfigurationBuilder and insert the configuration source. This is just an example:

```
// add new configuration source
builder.Configuration.Add(new MyCustomConfigurationSource
{
```

```
    SourceConfig = //configure whatever source
    Optional = false,
    ReloadOnChange = true
});
```

Usually, you would create an extension method to add the configuration source more easily, as illustrated here:

```
builder.Configuration.AddMyCustomSource("source", optional:
false, reloadOnChange: true);
```

A really detailed concrete example about how to create a custom configuration provider has been written by Andrew Lock. You can find this in the *Further reading* section of this chapter.

Summary

In most cases, you will not need to add a different configuration provider or create your own configuration provider, but it's good to know how to change it, just in case. Also, using a typed configuration is a nice way to read and provide the settings. In classic ASP.NET, we used a manually created façade to read the application settings in a typed manner. Now, this is automatically done by just providing a type. This type will be automatically instantiated, filled, and provided, via DI.

To learn more about customizing DI in ASP.NET Core 6.0, let's have a look at the next chapter.

Further reading

You can refer to the following source for more information:

- *Creating a custom ConfigurationProvider in ASP.NET Core to parse YAML*, *Andrew Lock*: https://andrewlock.net/creating-a-custom-iconfigurationprovider-in-asp-net-core-to-parse-yaml/

3
Customizing Dependency Injection

In this third chapter, we'll take a look at ASP.NET Core **dependency injection** (**DI**) and how to customize it to use a different DI container, if needed.

In this chapter, we will be covering the following topics:

- Using a different DI container
- Exploring the `ConfigureServices` method
- Using a different `ServiceProvider`
- Introducing Scrutor

The topics in this chapter refer to the hosting layer of the ASP.NET Core architecture:

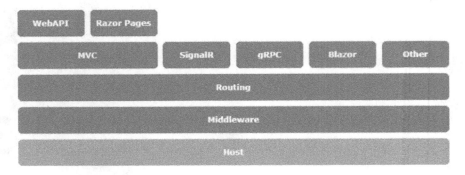

Figure 3.1 – ASP.NET Core architecture

Technical requirements

To follow the descriptions in this chapter, you will need to create an ASP.NET Core MVC application. Open your console, shell, or Bash terminal and change to your working directory. Use the following command to create a new MVC application:

```
dotnet new mvc -n DiSample -o DiSample
```

Now, open the project in Visual Studio by double-clicking the project file, or in Visual Studio Code by typing the following command in the already-open console:

```
cd DiSample
code .
```

All of the code samples in this chapter can be found in the GitHub repository for this book at https://github.com/PacktPublishing/Customizing-ASP. NET-Core-6.0-Second-Edition/tree/main/Chapter03.

Using a different DI container

In most projects, you don't really need to use a different DI container. The existing DI implementation in ASP.NET Core supports the main basic features and works both effectively and quickly. However, some other DI containers support a number of interesting features you might want to use in your application:

- Create an application that supports modules as lightweight dependencies using Ninject, for example, modules you might want to put into a specific directory and have them be automatically registered in your application.

- Configure the services in a configuration file outside the application, in an XML or JSON file instead of in C# only. This is a common feature in various DI containers, but not yet supported in ASP.NET Core.

- Add services at runtime, probably because you don't want to have an immutable DI container. This is also a common feature in some DI containers.

Let's now see how the `ConfigureServices` method enables you to use alternative DI containers.

Exploring the ConfigureServices method

Let's compare the current `ConfigureServices` method with a previous long-term support version to see what has changed. If you created a new ASP.NET Core project using version 3.1 and open `Startup.cs`, you will find the method to configure the services, which looks like this:

```
public void ConfigureServices(IServiceCollection services)
{
    services.Configure<CookiePolicyOptions>(options =>
    {
        // This lambda determines whether user
        // consent for non-essential cookies is
        // needed for a given request.
        options.CheckConsentNeeded = context => true;
    });

    services.AddControllersWithViews();
    services.AddRazorPages();
}
```

In contrast, in ASP.NET Core 6.0, there is no `Startup.cs` anymore, and the configuring of the services is done in `Program.cs`, which looks like this:

```
var builder = WebApplication.CreateBuilder(args);

// Add services to the container.
builder.Services.AddControllersWithViews();

var app = builder.Build();

// The rest of the file isn't relevant for this chapter
```

In both cases, the method gets `IServiceCollection`, which is already filled with a bunch of services that are required by ASP.NET Core. This service was added by the hosting services and parts of ASP.NET Core that are executed before the `ConfigureServices` method was called.

Inside the method, some more services are added. First, a configuration class that contains cookie policy options is added to `ServiceCollection`. After that, the `AddMvc()` method adds another bunch of services required by the MVC framework. So far, we have around 140 services registered to `IServiceCollection`. However, the service collection isn't the actual DI container.

The actual DI container is wrapped in the so-called **service provider**, which will be created out of the service collection. `IServiceCollection` has an extension method registered to create an `IServiceProvider` out of the service collection, which you can see in the following code snippet:

```
IServiceProvider provider = services.BuildServiceProvider()
```

`ServiceProvider` contains the immutable container that cannot be changed at runtime. With the default `ConfigureServices` method, `IServiceProvider` is created in the background after this method is called.

Next, we'll learn more about applying an alternative `ServiceProvider` as part of the DI customization process.

Using a different ServiceProvider

Changing to a different or custom DI container is relatively easy if the other container already supports ASP.NET Core. Usually, the other container will use `IServiceCollection` to feed its own container. The third-party DI containers move the already-registered services to the other container by looping over the collection:

1. Let's start by using `Autofac` as a third-party container. Type the following command into your command line to load the NuGet package:

   ```
   dotnet add package Autofac.Extensions.DependencyInjection
   ```

   `Autofac` is good for this because you are easily able to see what is happening here.

2. To register a custom IoC container, you need to register a different `IServiceProviderFactory`. In that case, you'll want to use `AutofacServiceProviderFactory` if you use `Autofac`. `IServiceProviderFactory` will create a `ServiceProvider` instance. The third-party container should provide one, if it supports ASP.NET Core.

   You should place this small extension method in `Program.cs` to register `AutofacServiceProviderFactory` with `IHostBuilder`:

   ```csharp
   using Autofac;
   using Autofac.Extensions.DependencyInjection;

   namespace DiSample;

   public static class IHostBuilderExtension
   {
       public static IHostBuilder
         UseAutofacServiceProviderFactory(
           this IHostBuilder hostbuilder)
       {
           hostbuilder.UseServiceProviderFactory
             <ContainerBuilder>(
           new AutofacServiceProviderFactory());
           return hostbuilder;
       }
   }
   ```

Don't forget to add using statements to `Autofac` and `Autofac.Extensions.DependencyInjection`.

3. To use this extension method, you can use `AutofacServiceProvider` in `Program.cs`:

```
var builder = WebApplication.CreateBuilder(args);

builder.Host.UseAutofacServiceProviderFactory();

// Add services to the container.
builder.Services.AddControllersWithViews();
```

This adds the `AutofacServiceProviderFactory` function to `IHostBuilder` and enables the `Autofac` IoC container. If you have this in place, you will use `Autofac` if you add services to `IServiceCollection` using the default way.

Introducing Scrutor

You don't always need to replace the existing .NET Core DI container to get and use some cool features. At the beginning of this chapter, I mentioned the autoregistration of services, which can be done with other DI containers. This can also be done with a nice NuGet package called **Scrutor** (`https://github.com/khellang/Scrutor`) by *Kristian Hellang* (`https://kristian.hellang.com`). Scrutor extends `IServiceCollection` to automatically register services with the .NET Core DI container.

> **Note**
>
> Andrew Lock has published a pretty detailed blog post relating to Scrutor. Rather than just repeating what he said, I suggest that you just go ahead and read that post to learn more about it: *Using Scrutor to automatically register your services with the ASP.NET Core DI container*, available at `https://andrewlock.net/using-scrutor-to-automatically-register-your-services-with-the-asp-net-core-di-container/`.

Summary

Using the approaches we have demonstrated in this chapter, you will be able to use any .NET Standard-compatible DI container to replace the existing one. If the container of your choice doesn't include `ServiceProvider`, create your own that implements `IServiceProvider` and uses the DI container inside. If the container of your choice doesn't provide a method to populate the registered services in the container, create your own method. Loop over the registered services and add them to the other container.

Actually, the last step sounds easy but can be a hard task, because you need to translate all the possible `IServiceCollection` registrations into registrations of the other container. The complexity of that task depends on the implementation details of the other DI container.

Anyway, you have the choice to use any DI container that is compatible with .NET Standard. You can change a lot of the default implementations in ASP.NET Core.

This is also something you can do with the default HTTPS behavior on Windows, which we will learn more about in the next chapter.

4

Configuring and Customizing HTTPS with Kestrel

In **ASP.NET Core**, **HTTPS** is on by default, and it is a first-class feature. On Windows, the certificate that is needed to enable HTTPS is loaded from the Windows certificate store. If you create a project on **Linux** or **Mac**, the certificate is loaded from a certificate file.

Even if you want to create a project to run it behind an **IIS** or an **NGINX** web server, HTTPS is enabled. Usually, you would manage the certificate on the IIS or NGINX web server in that case. Having HTTPS enabled here shouldn't be a problem, however, so don't disable it in the ASP.NET Core settings.

Managing the certificate within the ASP.NET Core application directly makes sense if you run services behind the firewall, services that are not accessible from the internet, services such as background services for a microservice-based application, or services in a self-hosted ASP.NET Core application.

There are also some scenarios on Windows where it makes sense to load the certificate from a file. This could be in an application that you will run on **Docker** for Windows or Linux. Personally, I like the flexible way of loading the certificate from a file.

Only two topics will be covered in this short chapter:

- Introducing Kestrel
- Setting up Kestrel

The topics in this chapter refer to the hosting layer of the ASP.NET Core architecture:

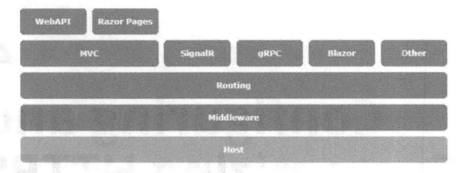

Figure 4.1 – The ASP.NET Core architecture

Technical requirements

To follow the descriptions in this chapter, you will need to create an ASP.NET Core MVC application. To do this, open your console, shell, or Bash terminal, and change to your working directory. Then, use the following command to create a new MVC application:

```
dotnet new mvc -n HttpSample -o HttpSample
```

Now, open the project in Visual Studio by double-clicking the project file, or in Visual Studio Code by typing the following command in the already-open console:

```
cd HttpSample
code .
```

All of the code samples in this chapter can be found in the **GitHub** repository for this book at https://github.com/PacktPublishing/Customizing-ASP. NET-Core-6.0-Second-Edition/tree/main/Chapter04.

Introducing Kestrel

Kestrel is a newly implemented HTTP server that is the hosting engine of ASP.NET Core. Every ASP.NET Core application will run on the Kestrel server. Classic ASP.NET applications (running on **.NET Framework**) usually run directly on the IIS. With ASP. NET Core, Microsoft was inspired by **Node.js**, which also ships an HTTP server called **libuv**. In the first version of ASP.NET Core, Microsoft also used libuv, and then it added a layer on top called Kestrel. At that time, Node.js and ASP.NET Core shared the same HTTP server.

Since the .NET Core framework has grown and **.NET sockets** have been implemented on it, Microsoft has built its own HTTP server based on .NET sockets and removed libuv, which was a dependency they don't own and control. Now, Kestrel is a full-featured HTTP server that runs ASP.NET Core applications.

The IIS acts as a reverse proxy that forwards the traffic to Kestrel and manages the Kestrel process. On Linux, usually NGINX is used as a reverse proxy for Kestrel.

Setting up Kestrel

As we did in the first two chapters of this book, we need to override the default `WebHostBuilder` a little bit to set up Kestrel. With ASP.NET Core 3.0 and later, it is possible to replace the default Kestrel base configuration with a custom configuration. This means that the Kestrel web server is configured to the host builder. Let's look at the steps to set up:

1. You will be able to add and configure Kestrel manually simply by using it. The following code shows what happens when you call the `UseKestrel()` method on `IwebHostBuilder`. Let's now see how this fits into the `CreateWebHostBuilder` method:

```
public class Program
{
    public static void Main(string[] args)
    {
        CreateWebHostBuilder(args).Build().Run();
    }

    public static IHostBuilder
        CreateHostBuilder(string[] args) =>
            Host.CreateDefaultBuilder(args)
```

```
            .ConfigureWebHostDefaults(webBuilder =>
        {
                webBuilder
                    .UseKestrel(options =>
                    {
                    })
                    .UseStartup<Startup>();
        }
}
```

The preceding code shows how the `Program.cs` looked until ASP.NET Core 5.0. In ASP.NET Core 6.0, the new minimal API approach is used to configure your application:

```
var builder = WebApplication.CreateBuilder(args);

builder.WebHost.UseKestrel(options =>
{
});

// Add services to the container.
builder.Services.AddControllersWithViews();

var app = builder.Build();

// the rest of this file is not relevant
```

We'll focus on the `UseKestrel()` method for the rest of this chapter. The `UseKestrel()` method accepts an action to configure the Kestrel web server.

2. What we *actually* need to do is configure the addresses and ports that the web server is listening on. For the HTTPS port, we also need to configure how the certificate should be loaded:

```
builder.WebHost.UseKestrel(options =>
{
    options.Listen(IPAddress.Loopback, 5000);
    options.Listen(IPAddress.Loopback,  5001,
        listenOptions =>
        {
```

```
        listenOptions.UseHttps("certificate.pfx",
            "topsecret");
    });
});
```

Don't forget to add a using statement to the `System.Net` namespace to resolve the `IPAddress`.

In this snippet, we add the addresses and ports to listen on. The configuration is defined as a secure endpoint configured to use HTTPS. The `UseHttps()` method is overloaded multiple times in order to load certificates from the Windows certificate store as well as from files. In this case, we will use a file called `certificate.pfx` located in the project folder.

3. To create a certificate file to just play around with this configuration, open the certificate store and export the development certificate created by Visual Studio. It is located in the current user certificates under the personal certificates:

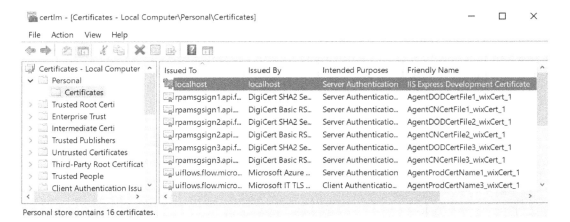

Figure 4.2 – Certificates

Right-click this entry. In the context menu, go to **All Tasks** and click **Export**. In the **Certificate Export Wizard**, click **Next** and then click **Yes, export the private key**, then click **Next**. Now, choose the **.PFX** format in the next screen and click **Next**. Here, you need to set a password. This is the exact same password you will need to use in the code, as seen in the following code example. Choose a filename and a location to store the file, and then click **Next**. The last screen will show a summary. Click **Finish** to save the certificate to a file.

For your safety

Use the following line *only* to play around with this configuration:

```
listenOptions.UseHttps("certificate.pfx", "topsecret");
```

To clarify why – the problem is the hardcoded password. Never, *ever* store a password in a code file that gets pushed to any source code repository. Ensure that you load the password from the configuration API of ASP.NET Core. Use the user secrets on your local development machine and use environment variables on a server. On **Azure**, use the application settings to store the passwords. Passwords will be hidden on the Azure portal UI if they are marked as passwords.

Summary

This is just a small customization, but it should help if you want to share the code between different platforms, or if you want to run your application on Docker and don't want to worry about certificate stores, and so on.

Usually, if you run your application behind a web server such as an IIS or NGINX, you don't need to care about certificates in your ASP.NET Core 6.0 application. However, if you host your application inside another application, on Docker, or without an IIS or NGINX, you will need to.

In the next chapter, we're going to talk about how to configure the hosting of ASP.NET Core web applications.

5

Configuring WebHostBuilder

When reading *Chapter 4, Configuring and Customizing HTTPS with Kestrel*, you might have asked yourself a question:

How can I use user secrets to pass the password to the HTTPS configuration?

You might even be wondering whether you can fetch the configuration from within `Program.cs`.

In this chapter, we're going to answer these questions through the following topic:

- Re-examining `WebHostBuilderContext`

The topic in this chapter refers to the hosting layer of the ASP.NET Core architecture:

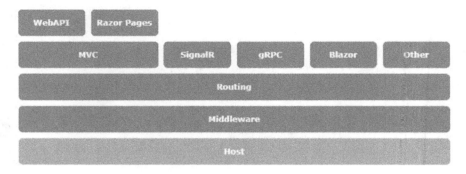

Figure 5.1 – The ASP.NET Core architecture

Technical requirements

To follow the descriptions in this chapter, you will need to create an **ASP.NET Core** application. To do this, open your console, shell, or Bash terminal, and change to your working directory. Then, use the following command to create a new web application:

```
dotnet new web -n HostBuilderConfig -o HostBuilderConfig
```

Now open the project in Visual Studio by double-clicking the project file, or in Visual Studio Code by typing the following command in the already-open console:

```
cd HostBuilderConfig
code .
```

> **Note**
>
> The simple web project template changed in .NET 6.0. In version 6.0, Microsoft introduced **minimal APIs** and changed the project template to use the minimal API approach. I'm going to show you the differences between these templates within this chapter.

All of the code samples in this chapter can be found in the **GitHub** repository for this book at https://github.com/PacktPublishing/Customizing-ASP. NET-Core-6.0-Second-Edition/tree/main/Chapter05.

Re-examining WebHostBuilderContext

Remember the `WebHostBuilder` Kestrel configuration in the `Program.cs` file that we looked at in *Chapter 4, Configuring and Customizing HTTPS with Kestrel*? In that chapter, we saw that you should use user secrets to configure the certificates password, as shown in the following code snippet:

```
public class Program
{
    public static void Main(string[] args)
    {
        CreateHostBuilder(args).Build().Run();
    }

    public static IHostBuilder CreateHostBuilder(
        string[] args) =>
        Host.CreateDefaultBuilder(args)
            .ConfigureWebHostDefaults(webBuilder =>
            {
                webBuilder
                    .UseKestrel(options =>
                    {
                        options.Listen(
                            IPAddress.Loopback,
                            5000);
                        options.Listen(
                            IPAddress.Loopback,
                            5001,
                            listenOptions =>
                            {
                                listenOptions.UseHttps(
                                    "certificate.pfx",
                                    "topsecret");
                            });
                    })
                    .UseStartup<Startup>();
```

```
        });
    }
}
```

This snippet is still valid for .NET 5.0 and prior versions, and it still is valid for almost all web projects in .NET 6.0. But it is not valid for the web project template we use in the technical requirements section. The Program.cs file for a minimal API looks like this:

```
var builder = WebApplication.CreateBuilder(args);

var app = builder.Build();

app.MapGet("/", () => "Hello World!");

app.Run();
```

This means the configuration we implemented in the last chapter would look like this within a minimal API:

```
using System.Net;
var builder = WebApplication.CreateBuilder(args);
builder.WebHost.UseKestrel(options =>
{
    options.Listen(IPAddress.Loopback, 5000);
    options.Listen(IPAddress.Loopback, 5001,
        listenOptions =>
        {
            listenOptions.UseHttps(
                "certificate.pfx",
                "topsecret");
        });
});
```

The following is valid for all project templates in .NET 6.0 and prior versions.

Usually, you are not able to fetch the configuration inside this code. You need to know that the UseKestrel() method is overloaded, as you can see here:

```
.UseKestrel((host, options) =>
{
```

```
    // ...
})
```

This first argument is a `WebHostBuilderContext` instance, which you can use to access the configuration. So, let's rewrite the lambda a little bit to use this context:

```
builder.WebHost.UseKestrel((host, options) =>
{
    var filename = host.Configuration.GetValue(
        "AppSettings:certfilename", "");
    var password = host.Configuration.GetValue(
        "AppSettings:certpassword", "");

    options.Listen(IPAddress.Loopback, 5000);
    options.Listen(IPAddress.Loopback, 5001,
        listenOptions =>
        {
            listenOptions.UseHttps(filename, password);
        });
});
```

In this example, we write the keys using the colon divider, as this is the way in which you need to read nested configurations from `appsettings.json`:

```
{
    "AppSettings": {
        "certfilename": "certificate.pfx",
        "certpassword": "topsecret"
    },
    "Logging": {
        "LogLevel": {
            "Default": "Warning"
        }
    },
    "AllowedHosts": "*"
}
```

> **Important Note**
>
> This is just a sample of how to read configurations to configure **Kestrel**. Please never, *ever* store any credentials inside the code. Use the following concepts instead.

You are also able to read from the user secrets store where the keys are set up with the following **.NET CLI** commands that you need to execute in the project folder:

```
dotnet user-secrets init
dotnet user-secrets set "AppSettings:certfilename"
  "certificate.pfx"
dotnet user-secrets set "AppSettings:certpassword"
  "topsecret"
```

This also applies to environment variables:

```
SET APPSETTINGS_CERTFILENAME=certificate.pfx
SET APPSETTINGS_CERTPASSWORD=topsecret
```

> **Important Note**
>
> Since the user secrets store is for local development only, you should pass credentials via environment variables to the application in production, or production-like applications. Also, the app settings configuration you will make in Azure will be passed as environment variables to your application.

So, how does this all work?

How does it work?

Do you remember the days back when you needed to set app configurations in the Startup.cs file in ASP.NET Core 1.0? They were configured in the constructor of the StartUp class and looked similar to this if you added user secrets:

```
var builder = new ConfigurationBuilder()
    .SetBasePath(env.ContentRootPath)
    .AddJsonFile("appsettings.json")
    .AddJsonFile($"appsettings.{env.EnvironmentName}.json",
      optional: true);

if (env.IsDevelopment())
```

```
{
    builder.AddUserSecrets();
}


builder.AddEnvironmentVariables();
Configuration = builder.Build();
```

This code is now wrapped inside the CreateDefaultBuilder method (as you can see on GitHub – refer to the *Further reading* section for details) and looks like this:

```
builder.ConfigureAppConfiguration((hostingContext, config)
    =>
{
    var env = hostingContext.HostingEnvironment;

    config
        .AddJsonFile(
            "appsettings.json",
            optional: true,
            reloadOnChange: true)
        .AddJsonFile(
            $"appsettings.{env.EnvironmentName}.json",
            optional:  true,
            reloadOnChange: true);

    if (env.IsDevelopment())
    {
        var appAssembly = Assembly.Load(
            new AssemblyName(env.ApplicationName));
        if (appAssembly != null)
        {
            config.AddUserSecrets(appAssembly,
                optional: true);
        }
    }

    config.AddEnvironmentVariables();
```

```
    if (args != null)
    {
        config.AddCommandLine(args);
    }
});
```

This is almost the same code, and it is one of the first things that gets executed when building the WebHost.

This needs to be one of the first things we set up because Kestrel is configurable via the app configuration. You can specify ports, URLs, and more by using environment variables or appsettings.json.

You can find these lines in WebHost.cs on GitHub:

```
builder.UseKestrel((builderContext, options) =>
    {
        options.Configure(
            builderContext.Configuration.GetSection("Kestrel"));
    })
```

This means that you are able to add these lines to appsettings.json to configure Kestrel endpoints:

```
"Kestrel": {
    "EndPoints": {
        "Http": {
            "Url": "http://localhost:5555"
        }
    }
}
```

Alternatively, environment variables such as the following can be used to configure endpoints:

```
SET KESTREL_ENDPOINTS_HTTP_URL=http://localhost:5555
```

Let's now recap everything we've covered in this chapter.

Summary

Inside `Program.cs`, you are able to make app configurations inside the lambdas of the configuration methods, provided you have access to the `WebHostBuilderContext`. This way, you can use all the configurations you like to configure `WebHostBuilder`.

In the next chapter, we are going to have a look at the hosting details. You will learn about different hosting models and how to host an ASP.NET Core application in different ways.

Further reading

The `WebHost.cs` file in the ASP.NET GitHub repository:
`https://github.com/aspnet/MetaPackages/blob/d417aacd7c0eff202f7860fe1e686aa5beeedad7/src/Microsoft.AspNetCore/WebHost.cs`.

6
Using Different Hosting Models

In this chapter, we will talk about how to customize hosting in ASP.NET Core. We will look into the hosting options and different kinds of hosting, and take a quick look at hosting on IIS. This chapter is just an overview. It is possible to go into much greater detail for each topic, but that would fill a complete book on its own!

In this chapter, we will be covering the following topics:

- Setting up `WebHostBuilder`
- Setting up Kestrel
- Setting up `HTTP.sys`
- Hosting on IIS
- Using Nginx or Apache on Linux

The topics in this chapter refer to the hosting layer of the ASP.NET Core architecture:

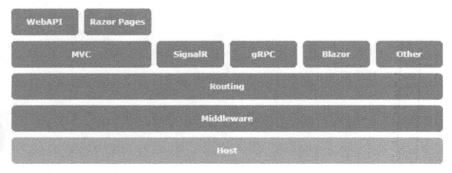

Figure 6.1 – ASP.NET Core architecture

This chapter tackles the following topics of the server architecture:

Figure 6.2 – ASP.NET server architecture

Technical requirements

For this chapter, we just need to set up a small, empty web application:

```
dotnet new web -n ExploreHosting -o ExploreHosting
```

That's it. Open it with Visual Studio Code:

```
cd ExploreHosting
code .
```

Et voilà! A simple project opens in Visual Studio Code.

The code for this chapter can be found on GitHub here: https://github.com/PacktPublishing/Customizing-ASP.NET-Core-6.0-Second-Edition/tree/main/Chapter06.

Setting up WebHostBuilder

As in the last chapter, we will focus on `Program.cs` in this section. `WebHostBuilder` is our friend. This is where we configure and create the web host.

The following code snippet is the default configuration of every new ASP.NET Core web project we create using **File | New | Project** in Visual Studio or the `dotnet new` command with the .NET CLI:

```
var builder = WebApplication.CreateBuilder(args);
var app = builder.Build();


app.MapGet("/", () => "Hello World!");


app.Run();
```

As we already know from previous chapters, the default builder has all the necessary stuff preconfigured. All you require in order to run an application successfully on Azure or an on-premises IIS is configured for you.

But you are able to override almost all of these default configurations, including the hosting configuration.

Next, let's set up Kestrel.

Setting up Kestrel

After `WebHostBuilder` is created, we can use various functions to configure the builder. Here, we can see one of them, which specifies the `Startup` class that should be used.

> **Note**
>
> As discussed in *Chapter 4, Configuring and Customizing HTTPS with Kestrel*, Kestrel is one possibility when it comes to hosting your application. Kestrel is a web server built into .NET and based on .NET socket implementations. Previously, it was built on top of **libuv**, which is the same web server that is used by Node.js. Microsoft removed the dependency to **libuv** and created their own web server implementation based on .NET sockets.

In the last chapter, we saw the `UseKestrel` method to configure the Kestrel options:

```
.UseKestrel((host, options) =>
{
```

```
    // ...
})
```

This first argument is `WebHostBuilderContext` to access already-configured hosting settings or the configuration itself. The second argument is an object to configure Kestrel. This code snippet shows what we did in the last chapter to configure the socket endpoints where the host needs to listen:

```
builder.WebHost.UseKestrel((host, options) =>
{
    var filename = host.Configuration.GetValue(
        "AppSettings:certfilename", "");
    var password = host.Configuration.GetValue(
        "AppSettings:certpassword", "");

    options.Listen(IPAddress.Loopback, 5000);
    options.Listen(IPAddress.Loopback,  5001,
        listenOptions  =>
        {
            listenOptions.UseHttps(filename, password);
        });
});
```

(You might need to add a `using` to `System.Net`.)

This will override the default configuration where you are able to pass in URLs, for example, using the `applicationUrl` property of `launchSettings.json` or an environment variable.

Let's now look at how to set up `HTTP.sys`.

Setting up HTTP.sys

There is another hosting option, a different web server implementation. `HTTP.sys` is a pretty mature library, deep within Windows, that can be used to host your ASP.NET Core application:

```
.UseHttpSys(options =>
{
    // ...
})
```

`HTTP.sys` is different from Kestrel. It cannot be used in IIS because it is not compatible with the ASP.NET Core module for IIS.

The main reason for using `HTTP.sys` instead of Kestrel is **Windows authentication**, which cannot be used in Kestrel. You can also use `HTTP.sys` if you need to expose your application to the internet without IIS.

> **Note**
>
> IIS has been running on top of `HTTP.sys` for years. This means that `UseHttpSys()` and IIS are using the same web server implementation. To learn more about `HTTP.sys`, please read the documentation, links to which can be found in the *Further reading* section.

Next, let's look at using IIS for hosting.

Hosting on IIS

An ASP.NET Core application shouldn't be directly exposed to the internet, even if it's supported for Kestrel or `HTTP.sys`. It would be best to have something such as a reverse proxy in between, or at least a service that watches the hosting process. For ASP.NET Core, IIS isn't just a reverse proxy. It also takes care of the hosting process, in case it breaks because of an error. If that happens, IIS will restart the process. Nginx may be used as a reverse proxy on Linux that also takes care of the hosting process.

> **Note**
>
> Be sure you created a new project or removed the Kestrel configuration of the previous section. This won't work with IIS.

To host an ASP.NET Core web on IIS or Azure, you need to publish it first. Publishing doesn't only compile the project; it also prepares the project for hosting on IIS, Azure, or a web server on Linux, such as Nginx.

The following command will publish the project:

```
dotnet publish -o ..\published -r win-x64
```

When viewed in a system browser, this should look as follows:

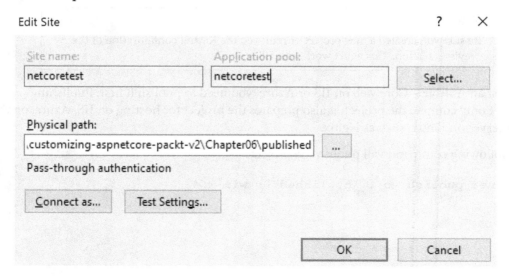

Figure 6.3 – A .NET published folder

This produces an output that can be mapped in IIS. It also creates a `web.config` to add settings for IIS or Azure. It contains the compiled web application as a DLL.

If you publish a self-contained application, it also contains the runtime itself. A self-contained application brings its own .NET Core runtime, but the size of the delivery increases a lot.

And on IIS? Just create a new web and map it to the folder where you placed the published output:

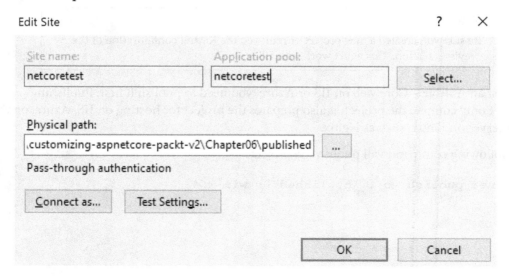

Figure 6.4 – The .NET publishing dialog

It gets a little more complicated if you need to change the security, if you have some database connections, and so on. This could be a topic for a separate chapter on its own.

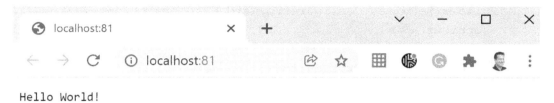

Figure 6.5 – Hello World! viewed in a browser

Figure 6.5 shows the output of the small `MapGet` in the `Program.cs` of the demo project:

```
app.MapGet("/", () => "Hello World!");
```

Next up, we'll discuss some alternatives for Linux.

Using Nginx or Apache on Linux

Publishing an ASP.NET Core application on Linux looks very similar to the way it looks on IIS, but preparing it for the reverse proxy requires some additional steps. You will need a web server such as Nginx or Apache as a reverse proxy that forwards the traffic to Kestrel and the ASP.NET Core application:

1. First, you need to allow your app to accept two specific forwarded headers. To do this, open `Startup.cs` and add the following lines to the `Configure` method before the `UseAuthentication` middleware:

```
app.UseForwardedHeaders(new ForwardedHeadersOptions
{
    ForwardedHeaders = ForwardedHeaders.XForwardedFor
        | ForwardedHeaders.XForwardedProto
});
```

2. You also need to trust the incoming traffic from the reverse proxy. This requires you to add the following lines to the ConfigureServices method:

```
Builder.Services.Configure<ForwardedHeadersOpticns>(
  options =>
{
    options.KnownProxies.Add(
        IPAddress.Parse("10.0.0.100"));
});
```

You might need to add a using to Microsoft.AspNetCore. HttpOverrides.

3. Add the IP address of the proxy here. This is just a sample.

4. Then, you need to publish the application:

```
dotnet publish --configuration Release
```

5. Copy the build output to a folder called /var/www/yourapplication. You should also do a quick test on Linux by calling the following command:

```
dotnet <yourapplication.dll>
```

6. Here, yourapplication.dll is the compiled application, including the path. If it is all working correctly, you should be able to call your web on http:// localhost:5000/.

    If it is working, the application should run as a service. This requires you to create a service file on /etc/systemd/system/. Call the file kestrel-yourapplication.service and place the following content in it:

```
[Unit]
Description=Example  .NET Web API App running on Ubuntu

[Service]
WorkingDirectory=/var/www/yourapplication
ExecStart=/usr/bin/dotnet/var/www/yourapplication/
yourapplication.dll
Restart=always

# Restart service after 10 seconds if the dotnet service
crashes:
```

```
RestartSec=10
KillSignal=SIGINT
SyslogIdentifier=dotnet-example
User=www-data
Environment=ASPNETCORE_ENVIRONMENT=Production
Environment=DOTNET_PRINT_TELEMETRY_MESSAGE=false

[Install]
WantedBy=multi-user.target
```

Ensure that the paths in lines 5 and 6 point to the folder where you placed the build output. This file defines that your app should run as a service on the default port. It also watches the app and restarts it in case it crashes. It also defines environment variables that get passed in to configure your application. See *Chapter 1*, *Customizing Logging*, to learn how to configure your application using environment variables.

Next up, we'll see how to configure Nginx.

Configuring Nginx

Now you can tell Nginx what to do using the following code:

```
server {
    listen          80;
    server_name     example.com *.example.com;
    location / {
        proxy_pass              http://localhost:5000;
        proxy_http_version 1.1;
        proxy_set_header    Upgrade $http_upgrade;
        proxy_set_header    Connection keep-alive;
        proxy_set_header    Host $host;
        proxy_cache_bypass $http_upgrade;
        proxy_set_header    X-Forwarded-For
                            $proxy_add_x_forwarded_for;
        proxy_set_header    X-Forwarded-Proto $scheme;
    }
}
```

This tells Nginx to forward calls on port 80 to example.com, and subdomains of it to http://localhost:5000, which is the default address of your application.

Configuring Apache

The Apache configuration looks pretty similar to the Nginx method, and does the same things at the end:

```
<VirtualHost *:*>
    RequestHeader set "X-Forwarded-Proto
      expr=%{REQUEST_SCHEME}
</VirtualHost>

<VirtualHost *:80>
    ProxyPreserveHost On
    ProxyPass / http://127.0.0.1:5000/
    ProxyPassReverse / http://127.0.0.1:5000/
    ServerName www.example.com
    ServerAlias *.example.com
    ErrorLog ${APACHE_LOG_DIR}yourapplication-error.log
    CustomLog ${APACHE_LOG_DIR}yourapplication-access.log
    common
</VirtualHost>
```

That's it for Nginx and Apache. Let's now wrap up this chapter.

Summary

ASP.NET Core and the .NET CLI already contain all the tools to get them running on various platforms and to set it up to get it ready for Azure and IIS, as well as Nginx. This is super easy and well described in the documentation.

Currently, we have WebHostBuilder, which creates the hosting environment of the applications. In version 3.0, we have HostBuilder, which is able to create a hosting environment that is completely independent of any web context.

ASP.NET Core 6.0 has a feature to run tasks in the background inside the application. To learn more about that, read the next chapter.

Further reading

For more information you can refer to the following links:

- **Kestrel documentation**: `https://docs.microsoft.com/en-us/aspnet/core/fundamentals/servers/kestrel?view=aspnetcore-6.0`

- **HTTP.sys documentation**: `https://docs.microsoft.com/en-us/aspnet/core/fundamentals/servers/httpsys?view=aspnetcore-6.0`

- **ASP.NET Core**: `https://docs.microsoft.com/en-us/aspnet/core/host-and-deploy/aspnet-core-module?view=aspnetcore-6.0`

7

Using IHostedService and BackgroundService

This seventh chapter isn't really about customization; it's more about a feature you can use to create background services to run tasks asynchronously inside your application. I use this feature to regularly fetch data from a remote service in a small ASP.NET Core application.

We'll examine the following topics:

- Introducing `IHostedService`
- Introducing `BackgroundService`
- Implementing the new Worker Service projects

The topics of this chapter refer to the Host layer of the ASP.NET Core architecture:

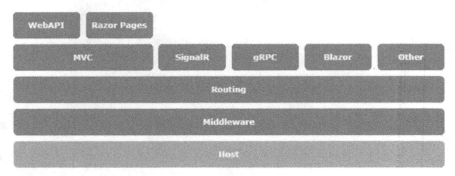

Figure 7.1 – ASP.NET Core architecture

Technical requirements

To follow the descriptions in this chapter, you will need to create an ASP.NET Core application. Open your console, shell, or Bash terminal, and change to your working directory. Use the following command to create a new MVC application:

```
dotnet new mvc -n HostedServiceSample -o HostedServiceSample
```

Now open the project in Visual Studio by double-clicking the project file or in VS Code by changing the folder to the project and typing the following command in the already open console:

```
cd HostedServiceSample
code .
```

All of the code samples in this chapter can be found in the GitHub repo for this book: https://github.com/PacktPublishing/Customizing-ASP.NET-Core-6.0-Second-Edition/tree/main/Chapter07.

Introducing IHostedService

Hosted services have been a thing since ASP.NET Core 2.0 and can be used to run tasks asynchronously in the background of your application. They can be used to fetch data periodically, do some calculations in the background, or do some cleanup. You can also use them to send preconfigured emails – or whatever you need to do in the background.

Hosted services are basically simple classes that implement the `IHostedService` interface. You call them with the following code:

```
public class SampleHostedService : IHostedService
{
    public Task StartAsync(CancellationToken
      cancellationToken)
    {
    }

    public Task StopAsync(CancellationToken
      cancellationToken)
    {
    }
}
```

`IHostedService` needs to implement a `StartAsync()` method and a `StopAsync()` method. The `StartAsync()` method is the place where you implement the logic to execute. This method gets executed once, immediately after the application starts. The `StopAsync()` method, on the other hand, gets executed just before the application stops. This also means that to start a scheduled service, you will need to implement it on your own. You will need to implement a loop that executes the code regularly.

To execute a `IHostedService`, you will need to register it in the ASP.NET Core dependency injection container as a singleton instance:

```
builder.Services.AddSingleton<IHostedService,
SampleHostedService>();
```

The next sample shows you how hosted services work. It writes a log message to the console on start, on stop, and every 2 seconds:

1. First, write the class skeleton that retrieves `ILogger` via `DependencyInjection`:

```
namespace HostedServiceSample;

public class SampleHostedService : IHostedService
{
    private readonly ILogger<SampleHostedService>
      logger;

    // inject a logger
    public
      SampleHostedService(ILogger<SampleHostedService>
        logger)
    {
        this.logger = logger;
    }

    public Task StartAsync(CancellationToken
      cancellationToken)
    {
    }

    public Task StopAsync(CancellationToken
      cancellationToken)
    {
    }
}
```

2. The next step is to implement the `StopAsync` method. This method is used to clean up in case you need to close connections, streams, and so on:

```
public Task StopAsync(CancellationToken
  cancellationToken)
{
    logger.LogInformation("Hosted service stopping");
```

```
        return Task.CompletedTask;
    }
```

3. The actual work will be done in the `StartAsync` method:

```
public Task StartAsync(CancellationToken
  cancellationToken)
{
    logger.LogInformation("Hosted service starting");

    return Task.Factory.StartNew(async () =>
    {
        // loop until a cancelation is requested
        while
          (!cancellationToken.IsCancellationRequested)
        {
            logger.LogInformation($"Hosted service
              executing - {DateTime.Now}");
            try
            {
                // wait for 2 seconds
                await
                  Task.Delay(TimeSpan.FromSeconds(2),
                    cancellationToken);
            }
            catch (OperationCanceledException) { }
        }
    }, cancellationToken);
}
```

4. To test this, start the application by calling the following command in the console:

```
dotnet run
```

Or press *F5* in Visual Studio or VS Code. This results in the following console output:

Figure 7.2 – A screenshot of the dotnet run output

As you can see, the log output is written to the console every 2 seconds.

In the next section, we will look at `BackgroundService`.

Introducing BackgroundService

The `BackgroundService` class was introduced in ASP.NET Core 3.0 and is basically an abstract class that implements the `IHostedService` interface. It also provides an abstract method, called `ExecuteAsync()`, which returns a `Task`.

If you want to reuse the hosted service from the last section, the code will need to be rewritten. Follow these steps to learn how:

1. First, write the class skeleton that retrieves `ILogger` via `DependencyInjection`:

```
namespace HostedServiceSample;

public class SampleBackgroundService :
  BackgroundService
{
    private readonly ILogger<SampleHostedService>
      logger;

    // inject a logger
    public SampleBackgroundService(
        ILogger<SampleHostedService> logger)
    {
        this.logger = logger;
    }
}
```

2. The next step would be to override the `StopAsync` method:

```
public override async Task StopAsync(CancellationToken
  cancellationToken)
{
    logger.LogInformation("Background service
      stopping");
    await Task.CompletedTask;
}
```

3. In the final step, we will override the `ExecuteAsync` method that does all the work:

```
protected override async Task
  ExecuteAsync(CancellationToken cancellationToken)
{
    logger.LogInformation("Background service
```

```
        starting");

    await Task.Factory.StartNew(async () =>
    {
        while
            (!cancellationToken.IsCancellationRequested)
        {
            logger.LogInformation($"Background service
                executing - {DateTime.Now}");
            try
            {
                await
                    Task.Delay(TimeSpan.FromSeconds(2),
                    cancellationToken);
            }
            catch (OperationCanceledException) {}
        }
    }, cancellationToken);
}
```

Even the registration is new.

Additionally, in ASP.NET Core 3.0 and later, the ServiceCollection has a new extension method to register hosted services or a background worker:

```
builder.Services.AddHostedService<SampleBackgroundService>();
```

To test this, start the application by calling the following command in the console:

```
dotnet run
```

Or press *F5* in Visual Studio or VS Code. It should show almost the same output as the SampleHostedService you created in the previous section.

Next, let's take a look at Worker Service projects.

Implementing the new Worker Service projects

The **worker services** and the generic hosting in ASP.NET Core 3.0 and later make it pretty easy to create simple service-like applications that can do some stuff without the full-blown ASP.NET stack – and without a web server.

You can create this project with the following command:

```
dotnet new worker -n BackgroundServiceSample -o
BackgroundServiceSample
```

Basically, this creates a console application with a `Program.cs` and a `Worker.cs` file in it. The `Worker.cs` file contains the `Worker` class that inherits from the `BackgroundService` class. In ASP.NET 5.0 and earlier, the `Program.cs` file looks pretty familiar to what we saw in the previous versions of ASP.NET Core but without a `WebHostBuilder`:

```
public class Program
{
    public static void Main(string[] args)
    {
        CreateHostBuilder(args).Build().Run();
    }


    public static IHostBuilder CreateHostBuilder(string[]
      args) =>
        Host.CreateDefaultBuilder(args)
            .ConfigureServices((hostContext, services) =>
            {
                services.AddHostedService<Worker>();
            });
}
```

In ASP.NET Core 6.0, `Program.cs` is pretty simplified in the same way as the minimal APIs. It looks like this:

```
using BackgroundServiceSample;

IHost host = Host.CreateDefaultBuilder(args)
    .ConfigureServices(services =>
    {
        services.AddHostedService<Worker>();
    })
    .Build();

await host.RunAsync();
```

This creates an `IHost` with dependency injection enabled. This means we can use dependency injection in any kind of .NET Core application, and not only in ASP.NET Core applications.

Then the worker is added to the service collection.

Where is this useful? You can run this app as a Windows service or as a background application in a Docker container, which doesn't need an HTTP endpoint.

Summary

You can now start to do some more complex things with an `IHostedService` and the `BackgroundService`. Be careful with background services because they all run in the same application; if you use too much CPU or memory, this could slow down your application.

For bigger applications, I would suggest running such tasks in a separate application that is specialized for executing background tasks: a separate Docker container, a `BackgroundWorker` on Azure, Azure Functions, or something like that. However, it should be separate from the main application in that case.

In the next chapter, we will learn about middleware, and how you can use them to implement special logic on the request pipeline or serve specific logic on different paths.

8

Writing Custom Middleware

Wow, we are already onto the eighth chapter of this book! In this chapter, we will learn about **middleware** and how you can use it to customize your app a little more. We will quickly go over the basics of middleware and then we'll explore some special things you can do with it.

In this chapter, we'll cover the following topics:

- Introducing middleware
- Writing custom middleware
- Exploring the potential of middleware
- Using middleware on ASP.NET Core 3.0 and later

The topics covered in this chapter relate to the middleware layer of the ASP.NET Core architecture:

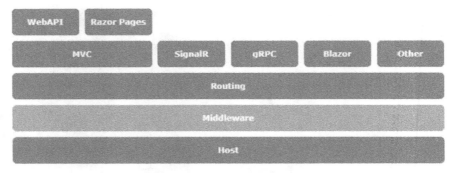

Figure 8.1 – The ASP.NET Core architecture

Technical requirements

To follow the descriptions in this chapter, you will need to create an ASP.NET Core MVC application. To do this, open your console, shell, or Bash terminal, and change to your working directory. Then, use the following command to create a new MVC application:

```
dotnet new web -n MiddlewaresSample -o MiddlewaresSample
```

Now, open the project in Visual Studio by double-clicking the project file, or in Visual Studio Code by typing the following command in the already-open console:

```
cd MiddlewaresSample
code .
```

> **Note**
>
> The simple web project template changed in .NET 6.0. In version 6.0, Microsoft introduced **minimal APIs** and changed the project template to use the minimal API approach. This is a simpler way to bootstrap and to get started with a web application.

All of the code samples in this chapter can be found in the **GitHub** repository for this book at https://github.com/PacktPublishing/Customizing-ASP.NET-Core-6.0-Second-Edition/tree/main/Chapter08.

Introducing middleware

The majority of you probably already know what middleware is, but some of you might not. Even if you have already been using ASP.NET Core for a while, you don't really need to know about middleware instances in detail, as they are mostly hidden behind nicely named extension methods such as `UseMvc()`, `UseAuthentication()`, `UseDeveloperExceptionPage()`, and so on. Every time you call a `Use` method in the `Startup.cs` file, in the `Configure` method, you'll implicitly use at least one – or maybe more – middleware components.

A middleware component is a piece of code that handles the request pipeline. Imagine the request pipeline as a huge tube where you can call something and then an echo comes back. The middleware is responsible for the creation of this echo – it manipulates the sound to enrich the information, handling the source sound, or handling the echo.

Middleware components are executed in the order in which they are configured. The first middleware component configured is the first that gets executed.

In an ASP.NET Core web application, if the client requests an image or any other static file, `StaticFileMiddleware` searches for that resource and returns that resource if it finds it. If not, this middleware component does nothing except call the next one. If there is no final piece of middleware that handles the request pipeline, the request returns nothing. The `MvcMiddleware` component also checks the requested resource, tries to map it to a configured route, executes the controller, creates a view, and returns an HTML or web API result. If `MvcMiddleware` does not find a matching controller, it will return a result anyway – in this case, it is a `404` status result. So, in any case, it returns an echo. This is why `MvcMiddleware` is the last piece of middleware configured.

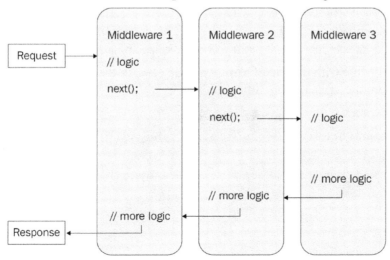

Figure 8.2 – A middleware workflow diagram

Exception-handling middleware is usually one of the first pieces of middleware configured, not because it is the first to be executed, but because it is the last. The first middleware component configured is also the last one to be executed if the echo comes back from the tube. Exception-handling middleware validates the result and displays a possible exception in a browser in a client-friendly way. The following process describes where a runtime error gets a 500 status:

1. You are able to see how the pipeline is executed if you create an empty ASP.NET Core application, as described in the *Technical requirements* section.

2. Open Program.cs with your favorite editor. This should be pretty small compared to a regular ASP.NET Core application. In ASP.NET Core 6.0, Microsoft introduced the minimal API approach, which simplifies the application configuration and hides a lot of default configuration from the developers. Microsoft is also implementing default using statements in ASP.NET Core. Because of this, you don't see any using statements initially. This is how Program.cs looks in ASP.NET Core 6.0:

```
var builder = WebApplication.CreateBuilder(args);
var app = builder.Build();


app.MapGet("/", () => "Hello World!");


app.Run();
```

Here, special lambda middleware is bound to the default route and only writes "Hello World!" to the response stream. The response stream is the echo we learned about previously. This special middleware stops the pipeline and returns something like an echo. So, it is the last middleware to run.

3. Replace the line with the call of app.MapGet() with the following lines of code, right before the app.Run() function:

```
app.Use(async (context, next) =>{
    await context.Response.WriteAsync("===");
    await next();
    await context.Response.WriteAsync("===");
});

app.Use(async (context, next) =>
{
    await context.Response.WriteAsync(">>>>>> ");
```

```
        await next();
        await context.Response.WriteAsync(" <<<<<<");
    });

    app.Run(async context =>
    {
        await context.Response.WriteAsync("Hello World!");
    });
```

These two calls of app.Use() also create two lambda middlewares, but this time, in addition to dealing with specific requests, the middleware components are calling their successors: each middleware component knows which middleware component should follow it, and so calls it. The call of app.Run() replaces the call of app.MapGet(), but it does basically the same thing, except that app.Run() directly writes to the response stream. The lambda middlewares created with app.Use() write to the response stream before and after the next middleware is called. Before the next middleware is called, the actual request is handled, and after the next middleware is called, the response (echo) is handled. This should demonstrate how the pipeline works.

4. If you now run the application (using dotnet run) and open the displayed URL in the browser, you should see a plain text result like this:

```
===>>>>>> Hello World! <<<<<<===
```

Does this make sense to you? If yes, let's move on and see how to use this concept to add some additional functionality to the request pipeline.

Writing custom middleware

ASP.NET Core is based on middleware. All the logic that gets executed during a request is based on middleware. So, we can use this to add custom functionality to the web. In the following process, we want to find out the execution time of every request that goes through the request pipeline:

1. We can do this by creating and starting a stopwatch before the next middleware is called, and stop measuring the execution time after the next middleware is called, like so:

```
app.Use(async (context, next) =>
{
```

```
var s = new Stopwatch();
s.Start();

// execute the rest of the pipeline
await next();

s.Stop(); //stop measuring
var result = s.ElapsedMilliseconds;

// write out the milliseconds needed
await context.Response.WriteAsync($" Time needed:
    {result} milliseconds");
});
```

You might need to add a `using` statement for `System.Diagnostics`.

After that, we return the elapsed milliseconds to the response stream.

2. If you write some more middleware components, the configuration in
    `Program.cs` gets pretty messy. This is why most middleware components are
    written as separate classes. This could look like this:

```
using System.Diagnostics;

public class StopwatchMiddleware
{
    private readonly RequestDelegate _next;

    public StopwatchMiddleware(RequestDelegate next)
    {
        _next = next;
    }

    public async Task Invoke(HttpContext context)
    {
        var s = new Stopwatch();
        s.Start();

        // execute the rest of the pipeline
```

```
            await _next(context);

            s.Stop(); //stop measuring
            var result = s.ElapsedMilliseconds;

            // write out the milliseconds needed
            await context.Response.WriteAsync($" Time
                needed: {result} milliseconds");
        }
    }
```

This way, we get the next middleware component to execute via the constructor and the current context in the Invoke() method.

> **Note**
>
> The middleware is initialized at the start of the application and the constructor runs only once during the application lifetime. On the other hand, the Invoke() method is called once per request.

3. To use this middleware, there is a generic UseMiddleware() method available that you can use:

```
app.UseMiddleware<StopwatchMiddleware>();
```

4. However, the more elegant way is to create an extension method that encapsulates this call:

```
public static class StopwatchMiddlewareExtension
{
    public static IApplicationBuilder
      UseStopwatch(this IApplicationBuilder app)
    {
        app.UseMiddleware<StopwatchMiddleware>();
        return app;
    }
}
```

5. Now, you can simply call it like this:

```
app.UseStopwatch();
```

This way, you can provide additional functionality to an ASP.NET Core application through the request pipeline. You have the entire `HttpContext` available in your middleware. With this, you can manipulate the request or even the response using middleware.

For example, `AuthenticationMiddleware` tries to collect user information from the request. If it doesn't find any, it will ask for the information by sending a specific response back to the client. If it finds some information, it will add it to the request context and make it available to the entire application this way.

Exploring the potential of middleware

There are many other things you can do with middleware. For example, did you know that you can split the request pipeline into two or more pipelines? We'll look at how to do that and several other things in this section.

Branching the pipeline with /map

The next code snippet shows how to create branches of the request pipeline based on specific paths:

```
app.Map("/map1", app1 =>
{
    // some more Middleware

    app1.Run(async context =>
    {
        await context.Response.WriteAsync("Map Test 1");
    });
});

app.Map("/map2", app2 =>
{
    // some more Middleware

    app2.Run(async context =>
    {
        await context.Response.WriteAsync("Map Test 2");
    });
```

```
});
```

```
// some more Middleware
```

The /map1 path is a specific branch that continues the request pipeline inside – this is the same with the /map2 path. Both maps have their own middleware configurations inside. All other unspecified paths will follow the main branch.

Branching the pipeline with MapWhen()

There is also a MapWhen() method to branch the pipeline based on a condition, instead of a branch based on a path:

```
public void Configure(IApplicationBuilder app)
{
    app.MapWhen(
        context =>
            context.Request.Query.ContainsKey("branch"),
        app1 =>
        {
            // some more Middleware

            app1.Run(async context =>
            {
                await context.Response.WriteAsync(
                    "MapBranch Test");
            });
    });

    // some more Middleware

    app.Run(async context =>
    {
        await context.Response.WriteAsync(
            "Hello from non-Map delegate.");
    });
}
```

Next, we'll look at using middleware to create conditions.

Creating conditions with middleware

You can create conditions based on configuration values or, as shown here, based on properties of the request context. In the previous example, a query string property was used. You can use HTTP headers, form properties, or any other property of the request context.

You are also able to nest the maps to create child and grandchild branches if needed.

We can use Map() or MapWhen() to provide a special API or resource based on a specific path or a specific condition, respectively. The ASP.NET Core HealthCheck API works like this: first, it uses MapWhen() to specify the port to use, and then, it uses Map() to set the path for the HealthCheck API (or, it uses Map() if no port is specified). In the end, HealthCheckMiddleware is used. The following code is just an example to show what this looks like:

```
private static void UseHealthChecksCore(IApplicationBuilder
    app, PathString path, int? port, object[] args)
{
    if (port == null)
    {
        app.Map(path,
            b =>
                b.UseMiddleware<HealthCheckMiddleware>(args));
    }
    else
    {
        app.MapWhen(
            c => c.Connection.LocalPort == port,
            b0 => b0.Map(path,
            b1 =>
                b1.UseMiddleware<HealthCheckMiddleware>(args)
                )
        );
    };
}
```

Next, let's see how you should use terminating middleware components in newer versions of ASP.NET Core.

Using middleware in ASP.NET Core 3.0 and later

In ASP.NET Core 3.0 and later, there are two new kinds of middleware element, and they are called `UseRouting` and `UseEndpoints`:

```
public void Configure(IApplicationBuilder app,
    IWebHostEnvironment env)
{
    if (env.IsDevelopment())
    {
        app.UseDeveloperExceptionPage();
    }

    app.UseRouting();

    app.UseEndpoints(endpoints =>
    {
        endpoints.MapGet("/", async context =>
        {
            await context.Response.WriteAsync("Hello
                                              World!");
        });
    });
}
```

The first one is a middleware component that uses routing and the other one uses endpoints. So, what exactly are we looking at?

This is the new **endpoint routing**. Previously, routing was part of MVC, and it only worked with MVC, web APIs, and frameworks that are based on the MVC framework. In ASP.NET Core 3.0 and later, however, routing is no longer in the MVC framework. Now, MVC and the other frameworks are mapped to a specific route or endpoint. There are different kinds of endpoint definitions available.

In the preceding code snippet, a GET request is mapped to the page root URL. In the next code snippet, MVC is mapped to a route pattern, and **Razor Pages** are mapped to the Razor Pages-specific file structure-based routes:

```
app.UseEndpoints(endpoints =>
{
    endpoints.MapControllerRoute(
        name: "default",
        pattern: "{controller=Home}/{action=Index}/{id?}");
    endpoints.MapRazorPages();
});
```

There is no UseMvc() method anymore, even if it still exists and works on the IApplicationBuilder object level, to prevent the existing code from breaking. Now, there are new methods to activate ASP.NET Core features more granularly.

These are the most commonly used new Map methods for ASP.NET Core 5.0 or later:

- **Areas for MVC and web API**: endpoints. MapAreaControllerRoute(...);
- **MVC and web API**: endpoints.MapControllerRoute(...);
- **Blazor server-side**: endpoints.MapBlazorHub(...);
- **SignalR**: endpoints.MapHub(...);
- **Razor Pages**: endpoints.MapRazorPages(...);
- **Health checks**: endpoints.MapHealthChecks(...);

There are many more methods to define fallback endpoints, to map routes and HTTP methods to delegates, and for middleware components.

If you want to create middleware that works on all requests, such as StopWatchMiddleware, this will work as before on IApplicationBuilder. If you would like to write middleware to work on a specific path or route, you will need to create a Map method for it to map it to that route.

> **Important Note**
> It is no longer recommended to handle the route inside the middleware. Instead, you should use the new endpoint routing. With this approach, the middleware is a lot more generic, and it will work on multiple routes with a single configuration.

I recently wrote middleware to provide a **GraphQL** endpoint in an ASP.NET Core application. However, I had to rewrite it to follow the new ASP.NET Core routing. The old way would still have worked, but it would have handled the paths and routes separately from the new ASP.NET Core routing. Let's look at how to deal with those situations.

Rewriting terminating middleware to meet the current standards

If you have existing middleware that provides a different endpoint, you should change it to use the new endpoint routing:

1. As an example, let's create small, dummy middleware that writes an application status to a specific route. In this example, there is no custom route handling:

```
namespace MiddlewaresSample;


public class AppStatusMiddleware
{
    private readonly RequestDelegate _next;
    private readonly string _status;


    public AppStatusMiddleware(
        RequestDelegate next, string status)
    {
        _next = next;
        _status = status;
    }

    public async Task Invoke(HttpContext context)
    {
        await context.Response.WriteAsync(
            $"Hello {_status}!");
    }
}
```

The first thing we need to do is write an `extension` method on the `IEndpointRouteBuilder` object. This method has a route pattern as an optional argument and returns an `IEndpointConventionBuilder` object to enable **cross-origin resource sharing (CORS)**, authentication, or other conditions to the route.

2. Now, we should add an extension method to make it easier to use the middleware:

```
public static class MapAppStatusMiddlewareExtension
{
    public static IEndpointConventionBuilder
      MapAppStatus(
        this IEndpointRouteBuilder routes,
        string pattern = "/",
        string name = "World")
    {
        var pipeline = routes
            .CreateApplicationBuilder()
            .UseMiddleware<AppStatusMiddleware>(name)
            .Build();

        return routes.Map(pattern, pipeline)
            .WithDisplayName("AppStatusMiddleware");
    }
}
```

3. Once that is complete, we can use the `MapAppStatus` method to map it to a specific route:

```
app.UseRouting();
app.UseEndpoints(endpoints =>
{
    endpoints.MapGet("/", () => "Hello World!");
    endpoints.MapAppStatus("/status", "Status");
});
```

4. We can now call the route in the browser by entering the following address: `http://localhost:5000/status`.

We will learn more about endpoint routing and how to customize it in *Chapter 9, Working with Endpoint Routing*. For now, let's recap what we've learned about middleware.

Summary

Most of the ASP.NET Core features are based on middleware and in this chapter, you learned how middleware works and how to create your own middleware components to extend the ASP.NET framework. You also learned how to use the new routing to add routes to your own custom terminating middleware components.

In the next chapter, we will have a look at the new endpoint routing in ASP.NET Core, which allows you to create your own hosted endpoints in an easy and flexible way.

9
Working with Endpoint Routing

In this chapter, we will talk about the new endpoint routing in **ASP.NET Core**. We will learn what endpoint routing is, how it works, where it is used, and how you are able to create your own routes to your own endpoints.

In this chapter, we will be covering the following topics:

- Exploring endpoint routing
- Creating custom endpoints
- Creating a more complex endpoint

The topics in this chapter refer to the routing layer of the ASP.NET Core architecture:

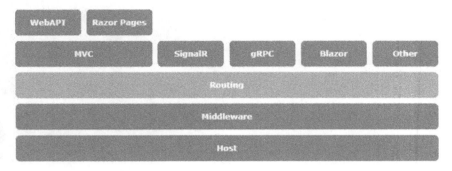

Figure 9.1 – The ASP.NET Core architecture

Technical requirements

For this series, we just need to set up a small, empty web application:

```
dotnet new mvc -n RoutingSample -o RoutingSample
```

That's it! Open the application with Visual Studio Code:

```
cd RoutingSample
code .
```

All of the code samples in this chapter can be found in the **GitHub** repository for this book at: https://github.com/PacktPublishing/Customizing-ASP. NET-Core-6.0-Second-Edition/tree/main/Chapter09.

Exploring endpoint routing

To learn about **endpoint routing**, you need to learn what an *endpoint* is and what *routing* is.

Endpoints are part of an app that get executed when a route maps the incoming request to it. Let's analyze this definition in a little more detail.

A client usually requests a resource from a server. In most cases, the client is a browser. The resource is defined by a URL, which points to a specific target. In most cases, the target is a web page. It could also be a mobile app that requests specific data from a JSON web API. What data the app requests is defined in the URL.

This means that the incoming request is also defined by the URL. The executing endpoint, on the other hand, is mapped to a specific route. A route is a URL or a pattern for a URL. ASP.NET Core developers are already familiar with such a route pattern:

```
app.UseRouting();
app.UseAuthorization();
app.UseEndpoints(endpoints =>
{
    endpoints.MapControllerRoute(
        name: "default",
        pattern: "{controller=Home}/{action=Index}/{id?}");
});
```

If the route or the route pattern matches the URL of the incoming request, the request gets mapped to that endpoint. In this case, the request gets mapped to the MVC endpoint.

ASP.NET Core can map to the following endpoints:

- Controllers (for example, MVC or web APIs)
- Razor Pages
- SignalR (and Blazor Server)
- gRPC services
- Health checks

Most of the endpoints have really simple route patterns. Only the MVC and web API endpoints use the more complex patterns. The route definitions of Razor pages are based on the folder and file structure of the actual pages.

Before endpoints were introduced in ASP.NET Core 2.2, routing was only a thing in MVC and web APIs. The implicit routing in **Razor Pages** was built-in there, and SignalR wasn't really ready. Blazor and gRPC weren't a thing back then, and the health checks were initially implemented as a middleware component.

Endpoint routing was introduced to separate routing from the actual endpoints. This makes the framework much more flexible, and it means that new endpoints don't need to implement their own kind of routing. This way, the endpoints can use the existing flexible routing technology to get mapped to a specific route.

Next, we'll see how you can create your own custom endpoints.

Creating custom endpoints

The easiest way to create an endpoint is by using the lambda-based endpoints:

```
app.Map("/map", async context =>
{
    await context.Response.WriteAsync("OK");
});
```

This maps the /map route to a simple endpoint that writes the word "OK" to the response stream.

> **A Note regarding Prior .NET 6.0 Versions**
>
> Prior to .NET 6.0, you would map custom endpoints on the endpoints object inside the lambda that gets passed to the UseEndpoints method in the Startup.cs file. With .NET 6.0 and the new **minimal API** approach, the mapping gets done on the app object in the Program.cs file.

You might need to add the Microsoft.AspNetCore.Http namespace to the using statements.

You can also map specific HTTP methods (such as GET, POST, PUT, and DELETE) to an endpoint. The following code shows how to map the GET and POST methods:

```
app.MapGet("/mapget", async context =>
{
    await context.Response.WriteAsync("Map GET");
});
app.MapPost("/mappost", async context =>
{
    await context.Response.WriteAsync("Map POST");
});
```

We can also map two or more HTTP methods to an endpoint:

```
app.MapMethods(
    "/mapmethods",
    new[] { "DELETE", "PUT" },
    async context =>
    {
```

```
      await context.Response.WriteAsync("Map Methods");
});
```

These endpoints look like the lambda-based terminating middleware components that we saw in *Chapter 8, Writing Custom Middleware*. These are middleware components that terminate the pipeline and return a result, such as HTML-based views, JSON structured data, or similar. Endpoint routing is a more flexible way to create an output, and it should be used in all versions from ASP.NET Core 3.0 onward.

In *Chapter 8, Writing Custom Middleware*, we saw that we can branch pipelines like this:

```
app.Map("/map", mapped =>
{
    // some more Middlewares
});
```

This also creates a route, but this will only listen to URLs that start with /map. If you would prefer to have a routing engine that handles patterns such as /map/{id:int?} to also match /map/456 and not /map/abc, you should use the new routing, as demonstrated earlier in this section.

Those lambda-based endpoints are useful for simple scenarios. However, because they are defined in Program.cs, things will quickly become messy if you start to implement more complex scenarios using this kind of lambda-based approach.

So, we should try to find a more structured way to create custom endpoints.

Creating a more complex endpoint

In this section, we will create a more complex endpoint, step by step. Let's do this by writing a really simple health check endpoint, similar to what you might need if you were to run your application inside a **Kubernetes** cluster, or just to tell others about your health status:

1. Microsoft advices starting with the definition of the API to add the endpoint from the developer's point of view. We do the same here. This means that we will add a MapSomething method first, without an actual implementation. This will be an extension method on the IEndpointRouteBuilder object. We are going to call it MapMyHealthChecks:

    ```
    // the new endpoint
    app.MapMyHealthChecks("/myhealth");
    ```

```
app.MapControllerRoute(
    name: "default",
    pattern:
        "{controller=Home}/{action=Index}/{id?}");
```

The new endpoint should be added in the same way as the prebuilt endpoints, so as not to confuse the developer who needs to use it.

Now that we know how the method should look, let's implement it.

2. Create a new static class called `MapMyHealthChecksExtensions` and place an extension method inside the `MapMyHealthChecks` object that extends `IEndpointRouteBuilder` and returns an `IEndpointConventionBuilder` object. I placed it in the `MapMyHealthChecksExtensions.cs` file:

```
namespace RoutingSample;

public static class MapMyHealthChecksExtensions
{
    public static IEndpointConventionBuilder
      MapMyHealthChecks (
        this IEndpointRouteBuilder endpoints,
        string pattern = "/myhealth")
    {
        // ...
    }
}
```

This is just the skeleton. Let's start with the actual endpoint first before using it.

3. The actual endpoint will be written as a *terminating* middleware component – that is, a middleware component that doesn't call the next one (see *Chapter 8*, *Writing Custom Middleware*) and creates an output to the response stream:

```
namespace RoutingSample;

public class MyHealthChecksMiddleware
{
    private readonly ILogger<MyHealthChecksMiddleware>
      _logger;
```

```
public MyHealthChecksMiddleware (
    RequestDelegate next,
    ILogger<MyHealthChecksMiddleware> logger)
{
    _logger = logger;
}

public async Task Invoke(HttpContext context)
{
    // add some checks here...
    context.Response.StatusCode = 200;
    context.Response.ContentType = "text/plain";
    await context.Response.WriteAsync("OK");
}
}
```

The actual work is done in the `Invoke` method. Currently, this doesn't really do more than respond with OK in plaintext and the 200 HTTP status, which is fine if you just want to show that your application is running. Feel free to extend the method with actual checks, such as checking for the availability of a database or related services, for example. Then, you would need to change the HTTP status and the output related to the result of your checks.

Let's use this terminating middleware.

4. Let's go back to the skeleton of the `MapMyHealthChecks` method. We now need to create our own pipeline, which we map to a given route. Place the following lines in that method:

```
var pipeline = endpoints
    .CreateApplicationBuilder()
    .UseMiddleware<MyHealthChecksMiddleware>()
    .Build();

return endpoints.Map(pattern, pipeline)
    .WithDisplayName("My custom health checks");
```

5.   This approach allows you to add some more middleware just for this new pipeline. The `WithDisplayName` extension method sets the configured display name to the endpoint.

6.   That's it! Press *F5* in your IDE to start the application and call `https://localhost:7111/myhealth` in your browser. You should see **OK** in your browser:

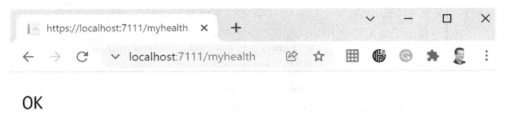

Figure 9.2 – A screenshot of the endpoint routing output

Please note the port number might vary. You can also convert an already existing terminating middleware component to a routed endpoint to benefit from much more flexible routing. And that's it for this chapter!

Summary

ASP.NET Core knows many ways in which to work with a request and to provide information to the requesting client. Endpoint routing is a way to provide resources based on the requested URL and the requested method.

In this chapter, you learned how to use a terminating middleware component as an endpoint that gets mapped to the new routing engine to be more flexible, matching the routes by which you want to serve the information to the requesting client.

Every web application needs to know its users to allow or restrict access to specific areas of the application or to specific data. In the next chapter, we show how to configure authentication to recognize your users.

10
Customizing ASP. NET Core Identity

In this tenth chapter, we are going to learn how to customize ASP.NET Core Identity. Security is one of the most important aspects of an application. Microsoft ships ASP.NET Core Identity as part of the ASP.NET Core framework to add authentication and user management to ASP.NET Core applications.

In this chapter, you will learn how to customize the basic implementation of the ASP.NET Core Identity UI and how to add custom information to **IdentityUser**. We'll cover the following points:

- Introducing ASP.Net Core Identity
- Customizing IdentityUser
- Customizing the Identity views

The topics of this chapter relate to the MVC layer of the ASP.NET Core architecture:

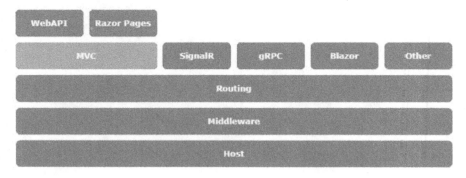

Figure 10.1 – ASP.NET Core architecture

Technical requirements

To follow the exercises in this chapter, you will need to create an ASP.NET Core MVC application. Open your console, shell, or Bash terminal and change to your working directory. Use the following command to create a new MVC application:

```
dotnet new mvc -n AuthSample -o AuthSample --auth Individual
```

Now, open the project in Visual Studio by double-clicking the project file, or open it in Visual Studio Code by typing the following command in the already-open console:

```
cd AuthSample
code .
```

All of the code samples for this chapter can be found in the GitHub repo for this book: https://github.com/PacktPublishing/Customizing-ASP.NET-Core-6.0-Second-Edition/tree/main/Chapter10.

Introducing ASP.NET Core Identity

An **identity** is basically an object that represents a user but could be a group as well. It is an object that helps you to know your user and their rights. An identity can have roles assigned that represent those rights. For example, a role called `writer` tells the application that the identity is allowed to write something. Identities can be nested as well. A user can be part of a group, and a group can be part of another group, and so forth.

ASP.NET Core Identity is a framework that structures this concept in .NET objects to help you store and read the user information. The framework also provides a mechanism to add a login form, a registration form, session handling, and so on. It also helps you to store the credentials in an encrypted and secure way.

ASP.NET Core Identity provides multiple ways to authenticate your users:

- **Individual**: The application manages the identities on its own. It has a database where user information is stored and manages the login, logout, registration, and so on on its own.

- **IndividualB2C**: Manages the user data on its own, but gets it from Azure B2C.

- **SingleOrg**: The identities get managed by Azure **Active Directory** (**AD**); the login, logout, and so on are done by Azure AD. The application just gets a ready-to-use identity from the web server.

- **MultiOrg**: Same as the previous but enabled for multiple Azure AD organizations.

- **Windows**: This means the classical Windows authentication, which is only available if the application is hosted with IIS. The user also gets a ready-to-use identity from the web server.

This chapter is not about the different ways to authenticate because this topic would fill an entire book.

Let's explore an application with **individual** authentication enabled.

As you might remember from the technical requirements, the --auth flag is used to create the application. It is set to Individual to create an ASP.NET Core MVC application with individual authentication enabled. This means it comes with a database to store the users. The --auth flag adds all the relevant code and dependencies to enable authentication in your freshly created application.

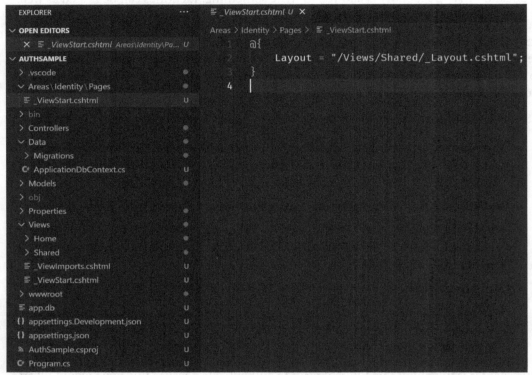

Figure 10.2 – Layout reference for the ASP.NET Identity UI

The --auth flag creates an area called Identity that contains a _ViewStart. cshtml file, which references the _Layout.cshtml file of the new project. The actual login or register screens are provided in a compiled library that is referenced to this project.

The AUTHSAMPLE contains a Data folder that contains an Entity Framework Core DbContext, as well as a database migration to create and update the database that is used here.

All the other parts, except Program.cs, are completely the same as in regular MVC applications.

If you created the application using the .NET CLI as shown in the technical requirements, a **SQLite** database is used. If you used Visual Studio to create this application, SQL Server is used to store the user data.

Before starting the application, call the following command in the terminal:

```
dotnet ef database update
```

This will create and update the database.

If it doesn't work, you might need to install the Entity Framework tool in the .NET CLI first:

```
dotnet tool install -g dotnet-ef
```

Then, call the following:

```
dotnet watch
```

The application will now start up in watch mode with hot-reload enabled. It will also open a browser window and call the application:

Figure 10.3 – AuthSample home page

As you can see, there is a menu on the upper right-hand side with the **Register** and **Login** options for this application. A click on the **Login** link takes you to the following login screen:

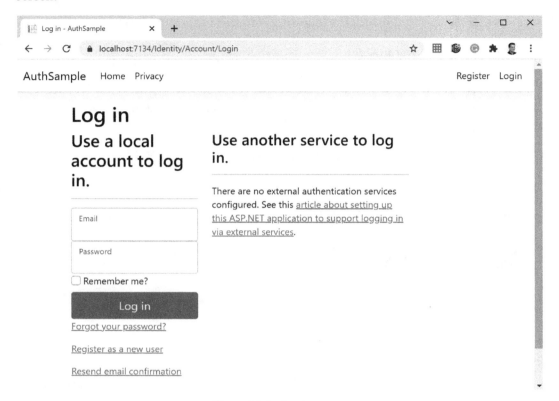

Figure 10.4 – Login screen

As mentioned, this view comes from a compiled Razor library that provides the necessary view to the Identity area. We automatically get this UI from the framework.

As the last thing in this section, we should have a quick look into Program.cs, which also differs from the files we saw in the last chapters.

In the upper section, where the services are registered, there are lines of code to register DbContext as well as database exception pages:

```
var connectionString = builder.Configuration
    .GetConnectionString("DefaultConnection");
builder.Services.AddDbContext<ApplicationDbContext>(
    options => options.UseSqlite(connectionString));
builder.Services.AddDatabaseDeveloperPageExceptionFilter();
```

```
builder.Services.AddDefaultIdentity<IdentityUser>(
    options => options.SignIn.RequireConfirmedAccount =
        true)
    .AddEntityFrameworkStores<ApplicationDbContext>();
```

There is also a registration for default identity that adds the `EntityFramework` store. It is also configured to only allow confirmed accounts, which means you as a user need to confirm your email address before you are allowed to log in.

In the lower section, where the middleware is used, we see that the authentication and authorization is used:

```
app.UseAuthentication();
app.UseAuthorization();
```

These two middleware enable authentication and authorization. The first tries to recognize the user by reading the authentication cookie. It also adds all the relevant information to the Identity object.

What you might need to do is to extend the user profile by adding some more properties to the user. Let's see how to do so in the next section.

Customizing IdentityUser

`IdentityUser` has the following fields: `Id`, `Username`, `Password`, `Email`, and `Phonenumber`.

Since the display name might differ from the username, we should add a `Name` property. Say we would like to send birthday wishes to the user; so, we would like to know their date of birth.

To do so, a file called `WebAppUser.cs` is added to the `Data` folder that contains the following lines:

```
using Microsoft.AspNetCore.Identity;

namespace AuthSample.Data;

public class WebAppUser : IdentityUser
{
    [PersonalData]
```

```
    public string? Name { get; set; }
    [PersonalData]
    public DateTime DOB { get; set; }
}
```

As shown here, WebAppUser derives from IdentityUser and extends it with the two already-mentioned properties.

In Program.cs, we need to modify the service registration to use the new WebAppUser:

```
builder.Services.AddDefaultIdentity<WebAppUser>
```

We also need to change DbContext in a way to use this WebAppUser, by changing the base class:

```
public class ApplicationDbContext :
    IdentityDbContext<WebAppUser, IdentityRole, string>
```

You might need to add a using statement to Microsoft.AspNetCore.Identity.

That's it for the first step. We now need to update the database:

```
dotnet ef migrations add CustomUserData
dotnet ef database update
```

Once you have IdentityUser extended with the custom properties, you can start to use this in the user profile. This needs some customization in the ASP.NET Core Identity UI.

Customizing the Identity views

Even if the ASP.NET Core Identity views come from a compiled Razor library, you can customize those views to add more fields or change the layout. To do so, you just need to override the given views with custom ones in the predefined folder structure within the area.

As mentioned, there is already an area called Identity in the project. Inside this area, there is a Pages folder. Here, a new folder called Account needs to be created, to match the route of the **Register** page.

If this is done, place a new Razor page called `Register.cshtml` inside this folder and put the following content inside just to see whether the overriding of views is working:

```
@page
@{
}

<h1>Hello Register Form</h1>
```

If you now run the app and click on **Register** in the upper left-hand corner, you will see the following page:

Figure 10.5 – Register page

It is working.

Actually, you don't need to override the views on your own. There is a code generator available to scaffold the views you'd like to override.

Install the code generator by calling this command:

```
dotnet tool install -g dotnet-aspnet-codegenerator
```

If not already done, you also need to install the following packages in your project:

```
dotnet add package Microsoft.VisualStudio.Web.CodeGeneration.Design
dotnet add package Microsoft.EntityFrameworkCore.Design
dotnet add package Microsoft.AspNetCore.Identity.EntityFrameworkCore
```

```
dotnet add package Microsoft.AspNetCore.Identity.UI
dotnet add package Microsoft.EntityFrameworkCore.SqlServer
dotnet add package Microsoft.EntityFrameworkCore.Tools
```

To learn what the code generator can do, run the following command:

```
dotnet aspnet-codegenerator identity -h
```

You can scaffold the entire Identity UI as well as specific pages. If you don't specify pages of the default UI, all pages will be generated in your project. To see which pages you can generate, type the following command:

```
dotnet aspnet-codegenerator identity -lf
```

The idea for the first change is to let the user fill in the name property on the registration page.

So, let's scaffold the **Register** page:

```
dotnet aspnet-codegenerator identity -dc AuthSample.Data.
ApplicationDbContext --files "Account.Register" -sqlite
```

This command tells the code generator to use the already-existing `ApplicationDbContext` and `Sqlite`. If you don't specify this, it will either create a new `DbContext` or register the existing `DbContext` to use with SQL Server instead of SQLite.

If all is done right, the code generator should only add the `Register.cshtml` page as well as some infrastructure files:

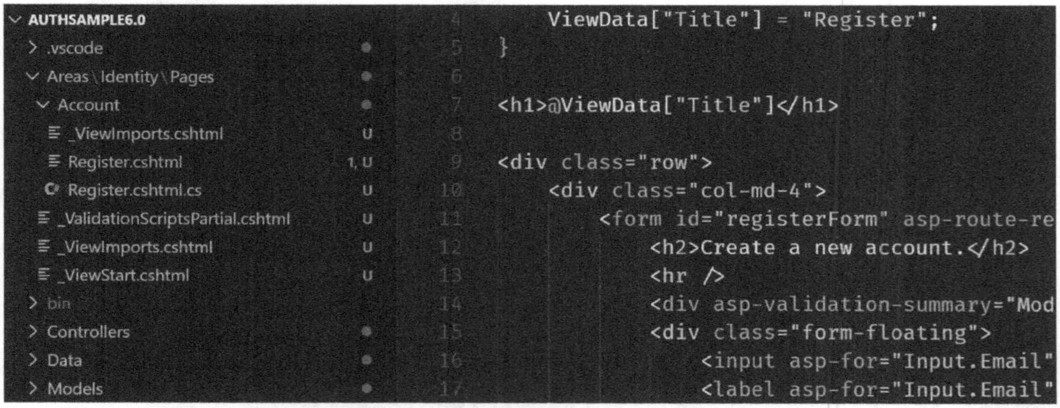

Figure 10.6 – Files added by the code generator

The code generator also knows that the project is using a custom `WebAppUser` instead of `IdentityUser`, which means `WebAppUser` is used in the generated code.

Now, you should change `Register.cshtml` to add the display name to the form. Add the following lines right before the form elements for the email field on line 15 and thereafter:

```
<div class="form-floating">
    <input asp-for="Input.Name" class="form-control"
        autocomplete="name" aria-required="true" />
    <label asp-for="Input.Name"></label>
    <span asp-validation-for="Input.Name"
        class="text-danger"></span>
</div>
```

Also, `Regiser.cshtml.cs` need to be changed. The `ImportModel` class needs the `Name` property:

```
public class InputModel
{
    [Required]
    [Display(Name = "Display name")]
    public string Name { get; set; }
```

In the `PostAsync` method, assign the `Name` property to the newly created user:

```
var user = CreateUser();
user.Name = Input.Name;
```

That's it.

After starting the application, you will see the following registration form:

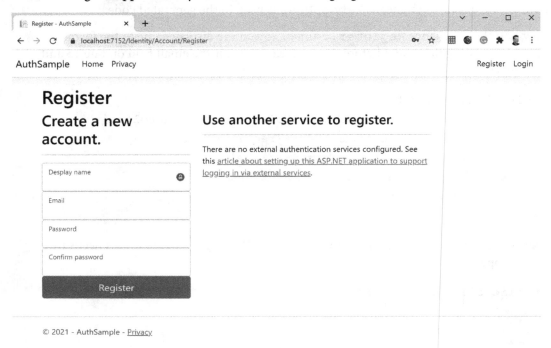

Figure 10.7 – Registration form

Try it out and you will see that it is working.

Since the user might need to update the name, we also need to change the view on the profile page. Here, the date of birth also needs to be added:

```
dotnet aspnet-codegenerator identity -dc AuthSample.Data.
ApplicationDbContext --files "Account.Manage.Index" -sqlite
```

Open the newly created `Index.cshtml.cs` that is located in the `/Areas/Identity/Pages/Account/Manage/` folder and place the following properties inside the `InputModel` class:

```
public class InputModel
{
    [Required]
    [Display(Name = "Display name")]
    public string Name { get; set; }
```

```
[Display(Name = "Date of birth")]
public DateTime DOB { get; set; }
```

You can now use these properties in the corresponding `Index.cshtml`. The next snippet needs to be placed between the validation summary and the username:

```html
<div class="form-floating">
    <input asp-for="Input.Name" class="form-control"
        autocomplete="name" aria-required="true" />
    <label asp-for="Input.Name"></label>
    <span asp-validation-for="Input.Name"
        class="text-danger"></span>
</div>
<div class="form-floating">
    <input asp-for="Input.DOB" class="form-control"
        type="date"/>
    <label asp-for="Input.DOB" class="form-label"></label>
</div>
```

This would be enough to display the fields, but there are some more changes needed to fill the form with saved data. Within the `LoadAsync` method, the instantiation of `InputModel` needs to be extended with the new properties:

```
Input = new InputModel
{
    PhoneNumber = phoneNumber,
    Name = user.Name,
    DOB = user.DOB
};
```

The changed values also need to get saved when the user saves the form. Place the next snippet right before the third-from-last line of the `OnPostAsync` method:

```
user.Name = Input.Name;
user.DOB = Input.DOB;
await _userManager.UpdateAsync(user);
```

This sets the values of `InputModel` to the `WebAppUser` properties and saves the changes in the database.

Let's try it out by calling `dotnet watch` in the terminal.

The profile page will now look similar to this:

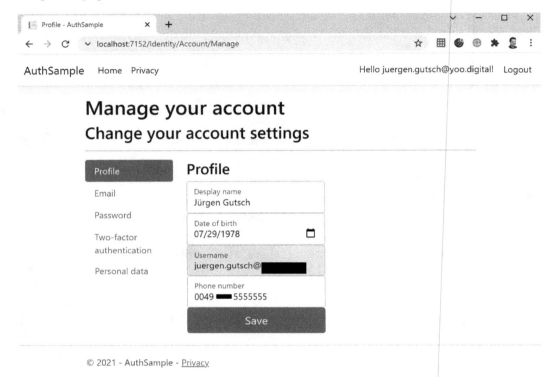

Figure 10.8 – Manage your account page

You can now change the display name and add your date of birth.

If the user provides a display name, they might show it in the upper left-hand corner after the login.

Open the `_LoginPartial.cshtml` that is in the `Views/Shared` folder and replace the first four lines with the following code snippet:

```
@using Microsoft.AspNetCore.Identity
@using AuthSample.Data
@inject SignInManager<WebAppUser> SignInManager
@inject UserManager<WebAppUser> UserManager
@{
  var user = await @UserManager.GetUserAsync(User);
}
```

This changes the generic type argument of `SignInManager` and `UserManager` from the `IdentityUser` type to the `WebAppUser` type. Inside the code block, the current `WebAppUser` is loaded via `UserManager` by passing the current user in.

Now, the output of the username on line 12 needs to be changed to write the display name:

```
Hello @user?.Name!
```

When `dotnet watch` is still running, the application running in the browser should already be updated. Maybe you need to log in again. You should now see the display name in the upper right-hand corner:

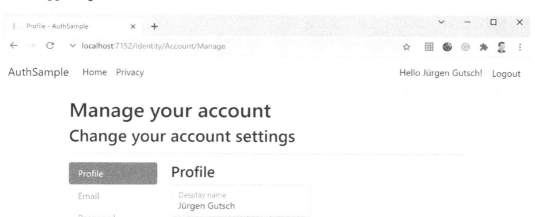

Figure 10.9 – Display name

That's it.

Summary

In this chapter, you learned how to extend ASP.NET Core Identity to enhance the user object by adding additional properties. You also learned how to enhance the Identity UI to load, save, and update the values of the new user properties.

But how would you manage roles for the users of your application?

This is what you will learn in the next chapter, about configuring identity management.

11
Configuring Identity Management

In the last chapter, we learned about how to add and customize the ASP.NET Core Identity UI to enable users to register, log in, and manage their profiles. Unfortunately, ASP.NET Core Identity doesn't provide identity management by default.

In this chapter, we are going to learn about how to manage ASP.NET Core Identity by using IdentityManager2 to create users and roles for your application.

We'll cover the following sections:

- Introducing IdentityManager2
- Setting up IdentityManager2

The topics of this chapter relate to the MVC layer of the ASP.NET Core architecture:

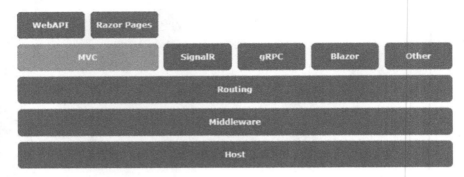

Figure 11.1 – ASP.NET Core architecture

Technical requirements

To follow the examples in this chapter, you will need to create an ASP.NET Core MVC application. Open your console, shell, or Bash terminal, and change to your working directory. Use the following command to create a new MVC application:

```
dotnet new mvc -n IdentityManagementSample -o
IdentityManagementSample --auth Individual
```

Now, open the project in Visual Studio by double-clicking the project file, or in VS Code by typing the following command in the already open console:

```
cd IdentityManagementSample
code .
```

All of the code samples of this chapter can be found in the GitHub repo for this book: https://github.com/PacktPublishing/Customizing-ASP.NET-Core-6.0-Second-Edition/tree/main/Chapter11.

> **Important Note**
> This chapter expects you to have completed the steps in the last chapter. As an alternative, you can reuse the project from the last chapter and might just need to adjust the project names.

Introducing IdentityManager2

IdentityManager is a project that was initially created and owned by Brock Allen who also created IdentityServer together with Dominick Baier. Scott Brady (https://www.scottbrady91.com/aspnet-identity) and his employer took over the project, ported it to ASP.NET Core, and released it as IdentityManager2 (https://brockallen.com/2018/07/09/identitymanager2/).

It is provided via NuGet (https://www.nuget.org/packages/identitymanager2).

Setting up IdentityManager2

The first step is to load the package. Use the already open command line or the terminals in VS Code or Visual Studio:

```
dotnet add package IdentityManager2
```

If the package is loaded, open `Program.cs` and add IdentityManager2 to the service collection:

```
builder.Services.AddIdentityManager();
```

Change the service registration of ASP.NET Identity from the following:

```
builder.Services.AddDefaultIdentity<ApplicationUser>{ …
```

To this:

```
builder.Services.AddIdentity<ApplicationUser, IdentityRole>( …
```

This adds some more relevant services to the service collection.

Also, `DefaultTokenProvider` needs to be added:

```
builder.Services.AddIdentity<ApplicationUser, IdentityRole>( …
)
    .AddEntityFrameworkStores<ApplicationDbContext>()
    .AddDefaultTokenProviders();
```

That's it with the services for now.

Then `IdentityServer` needs to be added to the pipeline. Add it after the authentications and authorization middleware:

```
app.UseRouting();

app.UseAuthentication();
app.UseAuthorization();

app.UseIdentityManager();

app.MapControllerRoute(
    name: "default",
    pattern: "{controller=Home}/{action=Index}/{id?}");
app.MapRazorPages();
```

Now we need to connect IdentityManager2 with the database connection that is already configured with ASP.NET Identity.

This needs the following package to be installed:

```
dotnet add package IdentityManager2.AspNetIdentity
```

Now you can connect IdentityManager2 with the already existing `ApplicationDbContext` that is `IdentityDbContext`, which handles `IdentityUsers` and `IdentityRoles`. Don't forget to add a `using` to `IdentityManager2.AspNetIdentity`. In the code, the already existing `ApplicationUser` needs to be used:

```
builder.Services.AddIdentityManager()
    .AddIdentityMangerService<
AspNetCoreIdentityManagerService<ApplicationUser,
 string, IdentityRole, string>>();
```

That's it to run `IdentityManager`. Type `dotnet watch` in Command Prompt to start the application or press *F5* in VS Code or VS. If you now call the application in the browser, you will see the UI to manage the data:

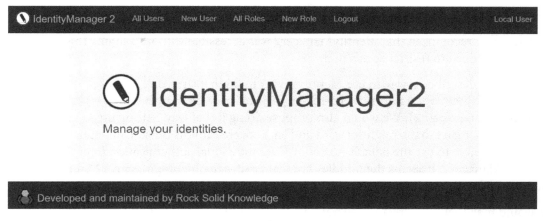

Figure 11.2 – IdentityManager2

Now you can create roles and users:

1. Create an **Admin** role and a **User** role. After that, create a User role for yourself.

2. After the User role is created, go to **All Users** and edit the newly created user. Here, you can change the user properties and assign both roles to them:

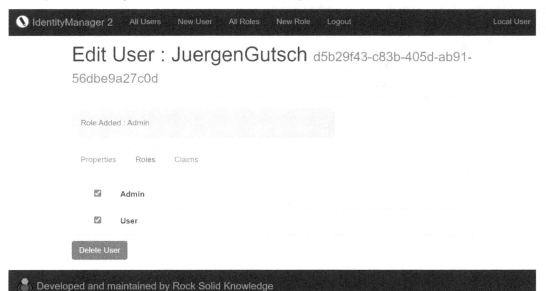

Figure 11.3 – Editing the roles

By using IdentityManager, you get a complete tool to manage your users and roles. It also works with custom users and custom user properties.

Securing IdentityManager2

I'm sure you recognized that IdentityManager2 was accessible without a login. This is by design. You need to restrict access to it.

Scott Brady described a way to use IdentityServer to do that (`https://www.scottbrady91.com/aspnet-identity/getting-started-with-identitymanager2`). We would also propose doing it that way. Setting up IdentityServer isn't that straightforward and isn't covered in this book. Unfortunately, it is not possible to use the default ASP.NET Core individual authentication to protect IdentityManager2. It seems the middleware that creates the IdentityManager2 UI doesn't support individual authentication and redirects to the ASP.NET Core Identity UI.

It would make sense to create a separate ASP.NET Core application that hosts IdentityManager2. This way, this kind of administrative UI would be completely separated from the publicly available application, and you would be able to use either OAuth or Azure Active Directory authentication to protect the application.

Summary

In this chapter, you learned how to add a user interface to manage the users and roles of your application. IdentityManager2 is the best and most complete solution to manage your identities.

In the next chapter, you will learn how to use content negotiation to create different kinds of outputs with only a single HTTP endpoint.

12

Content Negotiation Using a Custom OutputFormatter

In this chapter, we are going to learn about how to send your data to the client in different formats and types. By default, the ASP.NET Core web API sends data as JSON, but there are some more ways to distribute data.

We'll cover the following sections in this chapter:

- Introducing `OutputFormatter` objects
- Creating custom `Outputformatter` objects

The topics in this chapter relate to the WebAPI layer of the ASP.NET Core architecture:

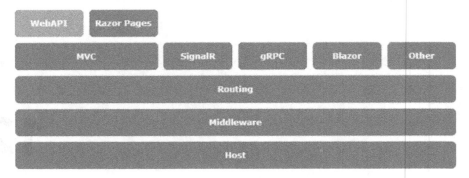

Figure 12.1 – The ASP.NET Core architecture

Technical requirements

To follow the examples in this chapter, you will need to create an ASP.NET Core MVC application. To do this, open your console, shell, or Bash terminal, and change to your working directory. Then, use the following command to create a new MVC application:

```
dotnet new webapi -n OutputFormatterSample -o
OutputFormatterSample
```

Now, open the project in Visual Studio by double-clicking the project file, or in Visual Studio Code by typing the following command in the already-open console:

```
cd OutputFormatterSample
code .
```

All of the code samples in this chapter can be found in the **GitHub** repository for this book at https://github.com/PacktPublishing/Customizing-ASP.NET-Core-6.0-Second-Edition/tree/main/Chapter12.

Introducing OutputFormatter objects

In ASP.NET Core, OutputFormatters are classes that transform your existing data into different formats to send it through HTTP to clients. The web API uses a default OutputFormatters to turn objects into JSON, which is the default format to send structured data. Other built-in formatters include an XML formatter and a plain text formatter.

With so-called *content negotiation*, clients are able to decide which format they want to retrieve. The client needs to specify the content type of the format in the `Accept` header. Content negotiation is implemented in `ObjectResult`.

By default, the web API always returns JSON, even if you accept text/XML in the header. This is why the built-in XML formatter is not registered by default.

There are two ways to add `XmlSerializerOutputFormatter` to ASP.NET Core:

- The first is shown in the following code snippet:

```
builder.Services.AddControllers()
    .AddXmlSerializerFormatters();
```

- Or, alternatively, you can use the following:

```
builder.Services.AddControllers()
    .AddMvcOptions(options =>
    {
        options.OutputFormatters.Add(
            new XmlSerializerOutputFormatter());
    });
```

You might need to add the `Microsoft.AspNetCore.Mvc.Formatters` namespace to the `using` statements.

There is also a formatter called `XmlDataContractSerializerOutputFormatter` available, which uses `DataContractSerializer` internally and is more flexible in its configurations.

By default, any `Accept` header will automatically be turned into `application/json`, even if you use one of these methods. However, we can fix that.

If you want to allow the clients to accept different headers, you need to switch that translation off:

```
builder.Services.AddControllers()
    .AddMvcOptions(options =>
    {
        options.RespectBrowserAcceptHeader = true;
        // false by default
    });
```

Some third-party components that don't completely support ASP.NET Core 5.0 or later won't write asynchronously to the response stream, but the default configuration since ASP.NET Core 3.0 *only* allows asynchronous writing.

To enable synchronous writing access, you will need to add these lines to the `ConfigureServices` method:

```
builder.Services.Configure<KestrelServerOptions>(options =>
{
    options.AllowSynchronousIO = true;
});
```

Add the `Microsoft.AspNetCore.Server.Kestrel.Core` namespace to the `using` statements to get access to the options.

To try the formatters, let's set up a small test project.

Preparing a test project

Using the console, we will create a small ASP.NET Core web API project, using the command shown previously in the *Technical requirements* section:

1. First, execute the following commands to add the necessary **NuGet** packages:

   ```
   dotnet add package GenFu
   dotnet add package CsvHelper
   ```

   This creates a new web API project and adds two NuGet packages to it: GenFu is an awesome library to easily create test data, and the second package, CsvHelper, helps us to easily write CSV data.

2. Now, open the project in Visual Studio or in VS Code and create a new API controller called `PersonsController` in the `controller` folder:

   ```
   [Route("api/[controller]")]
   [ApiController]
   public class PersonsController : ControllerBase
   {
   }
   ```

3. Open `PersonsController.cs` and add a `Get()` method like this:

   ```
   [HttpGet]
   public ActionResult<IEnumerable<Person>> Get()
   ```

```
{
    var persons = A.ListOf<Person>(25);
    return persons;
}
```

You might need to add the following using statements at the beginning of the file:

```
using GenFu;
using Microsoft.AspNetCore.Mvc;
using OutputFormatterSample.Models;
```

This creates a list of 25 persons by using GenFu. The properties will automatically be filled with realistic data. GenFu is an open source, fast, lightweight, and extendable test data generator. It contains built-in lists of names, cities, countries, phone numbers, and so on, and it fills the data automatically into the right properties of a class, depending on the property names. For example, a property called City will be filled with the name of a city, and a property called Phone, Telephone, or Phonenumber will be filled with a well-formatted fake phone number. You'll see the magic of GenFu and the results later on.

4. Now, create a Models folder, and create a new file called Person.cs with the Person class inside:

```
public class Person
{
    public int Id { get; set; }
    public string? FirstName { get; set; }
    public string? LastName { get; set; }
    public int Age { get; set; }
    public string? EmailAddress { get; set; }
    public string? Address { get; set; }
    public string? City { get; set; }
    public string? Phone { get; set; }
}
```

5. Open Program.cs as well, add the XML formatters, and allow other AcceptHeader, as described earlier:

```
builder.Services.AddControllers()
    .AddMvcOptions(options =>
    {
```

```
        options.RespectBrowserAcceptHeader = true;
            // false by default
        options.OutputFormatters.Add(
            new XmlSerializerOutputFormatter());
    });
```

That's it for now. Now, you are able to retrieve the data from the web API.

6. Start the project by using the `dotnet run` command.

Next, we'll test the API.

Testing the web API

The best tools to test a web API are **Fiddler** (`https://www.telerik.com/fiddler`) or **Postman** (`https://www.postman.com/`). I prefer Postman because I find it easier to use. You can use either tool, but in these demos, we will use Postman:

1. In Postman, create a new request. Enter the API URL, which is `https://localhost:5001/api/persons` (the port of the URL might vary), into the `address` field, and then, add a header with the `Accept` key and the `application/json` value.

2. After clicking **Send**, you will see the JSON result in the response body, as shown in the following screenshot:

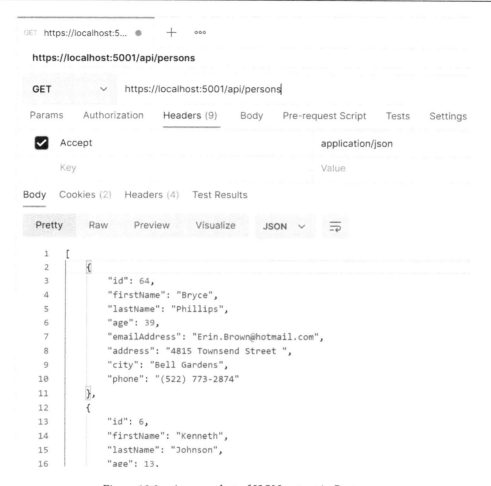

Figure 12.2 – A screenshot of JSON output in Postman

Here, you can see the autogenerated values. The GenFu object puts the data into the person's properties, based on the property type and the property name: real first names and real last names, as well as real cities and properly formatted phone numbers.

3.   Next, let's test the XML output formatter. In Postman, change the `Accept` header from `application/json` to `text/xml` and click **Send**:

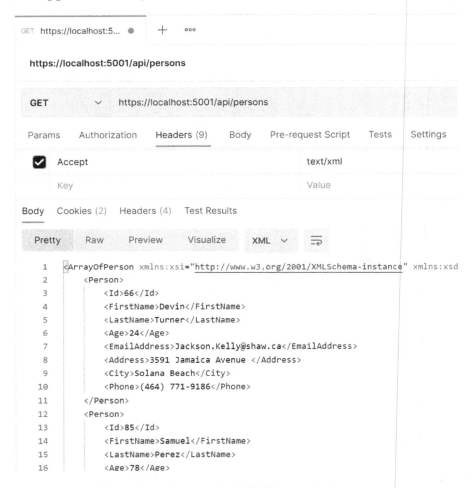

Figure 12.3 – A screenshot of XML output in Postman

We now have an XML-formatted output. Now, let's go a step further and create some custom `OutputFormatter` objects.

Creating custom OutputFormatter objects

In this example, our aim is to create a **VCard** output to be able to import the person's contact details directly into **Microsoft Outlook** or any other contact database that supports VCards. Later in this section, we also want to create a CSV output formatter.

Both are text-based output formatters, and they will derive their values from `TextOutputFormatter`. Let's look at the steps to create the VCard output:

1. Create a new class in a new file called `VcardOutputFormatter.cs`.

2. Now, insert the following class skeleton in the new file. You will find the implementations of the empty methods in the following code snippets. The constructor contains the supported media types and content encodings:

```
public class VcardOutputFormatter : TextOutputFormatter
{
    public string? ContentType { get; }

    public VcardOutputFormatter()
    {
        SupportedMediaTypes.Add(
            MediaTypeHeaderValue.Parse("text/vcard"));

        SupportedEncodings.Add(Encoding.UTF8);
        SupportedEncodings.Add(Encoding.Unicode);
    }

    protected override bool CanWriteType(Type type)
    {
    }

    public override Task WriteResponseBodyAsync(
        OutputFormatterWriteContext context,
        Encoding selectedEncoding)
    {
    }

    private static void FormatVcard(
        StringBuilder buffer,
        Person person,
        ILogger logger)
    {
```

```
    }
}
```

You might need to add the following using statements:

```
using Microsoft.AspNetCore.Mvc.Formatters;
using System.Text;
using Microsoft.Extensions.Logging;
using Microsoft.Net.Http.Headers;
using OutputFormatterSample.Models;
```

3. The next code snippet shows the implementation of the CanWriteType method. It is optional to override this method, but it makes sense to restrict it to a specific condition. In this case, the OutputFormatter can only format objects of the Person type:

```
protected override bool CanWriteType(Type type)
{
    if (typeof(Person).IsAssignableFrom(type)
        || typeof(IEnumerable<Person>)
            .IsAssignableFrom(type))
    {
        return base.CanWriteType(type);
    }
    return false;
}
```

4. You need to override WriteResponseBodyAsync to convert the actual Person objects into the output you want to have. To get the objects to convert, you need to extract them from OutputFormatterWriteContext object that gets passed into the method. You also get the HTTP response from this context. This is needed to write the results and send them to the client.

5. Inside the method, we check whether we get one person or a list of persons and call the not yet implemented FormatVcard method:

```
public override Task WriteResponseBodyAsync(
    OutputFormatterWriteContext context,
    Encoding selectedEncoding)
{
    var serviceProvider =
```

```
        context.HttpContext.RequestServices;
    var logger = serviceProvider.GetService(
        typeof(ILogger<VcardOutputFormatter>)) as
            ILogger;

    var response = context.HttpContext.Response;

    var buffer = new StringBuilder();
    if (context.Object is IEnumerable<Person>)
    {
        foreach (var person in context.Object as
            IEnumerable<Person>)
        {
            FormatVcard(buffer, person, logger);
        }
    }
    else
    {
        var person = context.Object as Person;
        FormatVcard(buffer, person, logger);
    }
    return response.WriteAsync(buffer.ToString());
}
```

6. To format the output to support standard `Vcard`, you need to do some manual work:

```
private static void FormatVcard(
    StringBuilder buffer,
    Person person,
    ILogger logger)
{
    buffer.AppendLine("BEGIN:VCARD");
    buffer.AppendLine("VERSION:2.1");
    buffer.AppendLine(
        $"FN:{person.FirstName} {person.LastName}");
    buffer.AppendLine(
```

```
        $"N:{person.LastName};{person.FirstName}");
    buffer.AppendLine(
        $"EMAIL:{person.EmailAddress}");
    buffer.AppendLine(
        $"TEL;TYPE=VOICE,HOME:{person.Phone}");
    buffer.AppendLine(
        $"ADR;TYPE=home:;;{person.Address};
            {person.City}");
    buffer.AppendLine($"UID:{person.Id}");
    buffer.AppendLine("END:VCARD");


    logger.LogInformation(
        $"Writing {person.FirstName}
            {person.LastName}");
}
```

7. Then, we need to register the new `VcardOutputFormatter` object in `Program.cs`:

```
builder.Services.AddControllers()
    .AddMvcOptions(options =>
    {
        options.RespectBrowserAcceptHeader = true;
        // false by default
        options.OutputFormatters.Add(
            new XmlSerializerOutputFormatter());

        // register the VcardOutputFormatter
        options.OutputFormatters.Add(
            new VcardOutputFormatter());
    });
```

You might need to add a `using` statement to `OutputFormatterSample`.

8. Start the app again using `dotnet run`.

9. Now, change the `Accept` header to `text/vcard`, and let's see what happens:

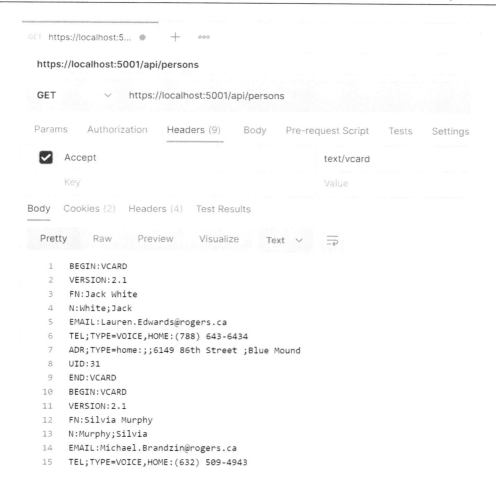

Figure 12.4 – A screenshot of VCard output in Postman

We now should see all of our data in VCard format.

10. Now, let's do the same for a CSV output. We already added the CsvHelper library to the project. So, go to the following URL and download CsvOutputFormatter to put it into your project: https://github.com/PacktPublishing/ Customizing-ASP.NET-Core-6.0-Second-Edition/blob/main/ Chapter12/OutputFormatterSample6.0/CsvOutputFormatter.cs

11. Let's have a quick look at the `WriteResponseBodyAsync` method:

```
public override async Task WriteResponseBodyAsync(
    OutputFormatterWriteContext context,
    Encoding selectedEncoding)
{
    var response = context.HttpContext.Response;

    var csv = new CsvWriter(
        new StreamWriter(response.Body),
        CultureInfo.InvariantCulture);
    IEnumerable<Person> persons;
    if (context.Object is IEnumerable<Person>)
    {
        persons = context.Object as
            IEnumerable<Person>;
    }
    else
    {
        var person = context.Object as Person;
        persons = new List<Person> { person };
    }
    await csv.WriteRecordsAsync(persons);
}
```

12. This almost works the same way as `VcardOutputFormatter`. We can pass the response stream via `StreamWriter` directly into `CsvWriter`. After that, we are able to feed the persons or the list of persons to the writer. That's it.

13. We also need to register `CsvOutputFormatter` before we can test it:

```
builder.Services.AddControllers()
    .AddMvcOptions(options =>
    {
        options.RespectBrowserAcceptHeader = true;
            // false by default
        options.OutputFormatters.Add(
            new XmlSerializerOutputFormatter());
```

```
// register the VcardOutputFormatter
options.OutputFormatters.Add(
    new VcardOutputFormatter());
// register the CsvOutputFormatter
options.OutputFormatters.Add(
    new CsvOutputFormatter());
});
```

14. In Postman, change the `Accept` header to `text/csv` and click **Send** again:

Figure 12.5 – A screenshot of text/CSV output in Postman

There we go – Postman was able to open all of the formats we tested.

Summary

Isn't that cool? The ability to change the format based on the `Accept` header is very handy. This way, you are able to create a web API for many different clients – an API that accepts many different formats, depending on the clients' preferences. There are still many potential clients out there that don't use JSON and prefer XML or CSV.

The other way around would be an option to consume CSV or any other format inside the web API. For example, let's assume your client sends you a list of people in CSV format. How would you solve this? Parsing the string manually in the `action` method would work, but it's not an easy option.

This is what `ModelBinder` objects can do for us. Let's see how they work in the next chapter.

13
Managing Inputs with Custom ModelBinder

In the last chapter regarding OutputFormatter, we learned about sending data out to clients in different formats. In this chapter, we are going to do it the other way. This chapter is about data you get in your web API from outside; for instance, what to do if you get data in a special format, or if you get data you need to validate in a special way. **Model Binders** will help you to handle this.

In this chapter, we will be covering the following topics:

- Introducing ModelBinder
- Preparing the test project
- Creating PersonsCsvBinder
- Using ModelBinder
- Testing ModelBinder

The topics in this chapter refer to the WebAPI layer of the ASP.NET Core architecture:

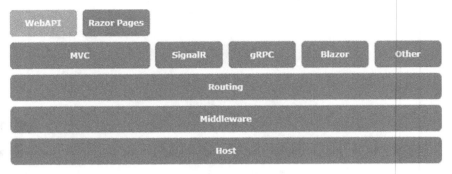

Figure 13.1 – ASP.NET Core architecture

Technical requirements

To follow the descriptions in this chapter, you will need to create an ASP.NET Core MVC application. Open your console, shell, or Bash terminal, and change to your working directory. Use the following command to create a new MVC application:

```
dotnet new webapi -n ModelBinderSample -o ModelBinderSample
```

Now, open the project in Visual Studio by double-clicking the project file or, in VS Code, by typing the following command in the already open console:

```
cd ModelBinderSample
code .
```

All of the code samples in this chapter can be found in the GitHub repository for this book at: https://github.com/PacktPublishing/Customizing-ASP. NET-Core-6.0-Second-Edition/tree/main/Chapter13.

Introducing ModelBinder

Model Binders are responsible for binding the incoming data to specific action method parameters. They bind the data sent with the request to the parameters. The default binders are able to bind data that is sent via the QueryString, or sent within the request body. Within the body, the data can be sent in URL or JSON format.

The model binding tries to find the values in the request by parameter names. The form values, route data, and query string values are stored as a key-value pair collection and the binding tries to find the parameter name in the keys of the collection.

Let's demonstrate how this works with a test project.

Preparing the test data

In this section, we're going to see how to send CSV data to a web API method. We will reuse the CSV data we created in *Chapter 12*, *Content Negotiation Using a Custom OutputFormatter*.

This is a snippet of the test data we want to use:

```
Id,FirstName,LastName,Age,EmailAddress,Address,City,Phone
48,Austin,Ward,49,Jake.Timms@live.com,"8814 Gravesend Neck Road
",Daly City,(620) 260-4410
2,Sierra,Smith,15,Elizabeth.Wright@hotmail.com,"1199 Marshall
Street ",Whittier,(655) 379-4362
27,Victorina,Radcliff,40,Bryce.Sanders@rogers.ca,"2663 Sutton
Street ",Bloomington,(255) 365-0521
78,Melissa,Brandzin,39,Devin.Wright@telus.net,"7439 Knight
Court ",Tool,(645) 343-2144
89,Kathryn,Perry,87,Hailey.Jenkins@hotmail.com,"5283 Vanderbilt
Street ",Carlsbad,(747) 369-4849
```

You can find the full CSV test data on GitHub at: `https://github.com/PacktPublishing/Customizing-ASP.NET-Core-6.0-Second-Edition/blob/main/Chapter13/testdata.csv`.

Preparing the test project

Let's prepare the project by following these steps:

1. In the already created project (refer to the *Technical requirements* section), we will now create a new empty API controller with a small action inside:

```
namespace ModelBinderSample.Controllers
{
    [Route("[controller]")]
    [ApiController]
    public class PersonsController : ControllerBase
```

```
    {
        public ActionResult<object> Post(
            IEnumerable<Person> persons)
        {
            return new
            {
                ItemsRead = persons.Count(),
                Persons = persons
            };
        }
    }
}
```

This looks basically like any other action. It accepts a list of persons and returns an anonymous object that contains the number of persons as well as the list of persons. This action is pretty useless but helps us to debug ModelBinder using Postman.

2. We also need the Person class:

```
public class Person
{
    public int Id { get; set; }
    public string? FirstName { get; set; }
    public string? LastName { get; set; }
    public int Age { get; set; }
    public string? EmailAddress { get; set; }
    public string? Address { get; set; }
    public string? City { get; set; }
    public string? Phone { get; set; }
}
```

This will actually work fine if we want to send JSON-based data to that action.

3. As a last preparation step, we need to add the CsvHelper NuGet package to parse the CSV data more easily. The .NET CLI is also useful here:

```
dotnet add package CsvHelper
dotnet add package System.Linq.Async
```

> **Note**
>
> The `System.Linq.Async` package is needed to handle the `IAsyncEnumerable` that gets returned by the `GetRecordsAsync()` method.

Now that this is all set up, we can try it out and create `PersonsCsvBinder` in the next section.

Creating PersonsCsvBinder

Let's build a binder.

To create `ModelBinder`, add a new class called `PersonsCsvBinder`, which implements `IModelBinder`. In the `BindModelAsync` method, we get `ModelBindingContext` with all the information in it that we need in order to get the data and deserialize it. The following code snippets show a generic binder that should work with any list of models. We have split it into sections so that you can clearly see how each part of the binder works:

```
public class PersonsCsvBinder : IModelBinder
{
    public async Task BindModelAsync(
        ModelBindingContext bindingContext)
    {
        if (bindingContext == null)
        {
            return;
        }

        var modelName = bindingContext.ModelName;
        if (String.IsNullOrEmpty(modelName))
        {
            modelName = bindingContext.OriginalModelName;
        }
        if (String.IsNullOrEmpty(modelName))
        {
            return;
        }
```

As you can see from the preceding code block, first, the context is checked against null. After that, we set a default argument name to the model, if none have already been specified. If this is done, we are able to fetch the value by the name we set previously:

```
var valueProviderResult =
    bindingContext.ValueProvider.GetValue(modelName);
if (valueProviderResult == ValueProviderResult.None)
{
    return;
}
```

In the next part, if there's no value, we shouldn't throw an exception in this case. The reason is that the next configured `ModelBinder` might be responsible. If we throw an exception, the execution of the current request is canceled and the next configured `ModelBinder` doesn't have the opportunity to be executed:

```
bindingContext.ModelState.SetModelValue(
    modelName, valueProviderResult);

var value = valueProviderResult.FirstValue;
// Check if the argument value is null or empty
if (String.IsNullOrEmpty(value))
{
    return;
}
```

If we have the value, we can instantiate a new `StringReader` that needs to be passed to `CsvReader`:

```
var stringReader = new StringReader(value);
var reader = new CsvReader(
    stringReader, CultureInfo.InvariantCulture);
```

With `CsvReader`, we can deserialize the CSV string value into a list of `Persons`. If we have the list, we need to create a new, successful `ModelBindingResult` that needs to be assigned to the `Result` property of `ModelBindingContext`:

```
        var asyncModel = reader.GetRecordsAsync<Person>();
        var model = await asyncModel.ToListAsync();
        bindingContext.Result =
            ModelBindingResult.Success(model);
    }
}
```

You might need to add the following `using` statements at the beginning of the file:

```
using Microsoft.AspNetCore.Mvc.ModelBinding;
using System.IO;
using CsvHelper
using System.Globalization;
```

Next, we'll put `ModelBinder` to work.

Using ModelBinder

The binder isn't used automatically because it isn't registered in the dependency injection container and is not configured to be used within the MVC framework.

The easiest way to use this model binder is to use `ModelBinderAttribute` on the argument of the action where the model should be bound:

```
[HttpPost]
public ActionResult<object> Post(
    [ModelBinder(binderType: typeof(PersonsCsvBinder))]
    IEnumerable<Person> persons)
{
    return new
    {
        ItemsRead = persons.Count(),
        Persons = persons
    };
}
```

Here, the type of our `PersonsCsvBinder` is set as `binderType` to that attribute.

> **Note**
> **Steve Gordon** wrote about a second option in his blog post, *Custom ModelBinding in ASP.NET MVC Core*. He uses a `ModelBinderProvider` to add the `ModelBinder` to the list of existing ones.

I personally prefer the explicit declaration because most custom `ModelBinder` will be specific to an action or to a specific type, and it prevents hidden magic in the background.

Now, let's test out what we've built.

Testing ModelBinder

To test it, we need to create a new request in Postman:

1. Start the application by running `dotnet run` in the console or by pressing *F5* in Visual Studio or VS Code.

2. In Postman, we will then set the request type to **POST** and insert the URL `https://localhost:5001/api/persons` in the address bar.

    The port number might vary on your side.

3. Next, we need to add the CSV data to the body of the request. Select `form-data` as the body type, add the `persons` key, and paste the following value lines in the value field:

```
Id,FirstName,LastName,Age,EmailAddress,Address,City,Phone
48,Austin,Ward,49,Jake.Timms@live.com,"8814 Gravesend
Neck Road ",Daly City,(620) 260-4410
2,Sierra,Smith,15,Elizabeth.Wright@hotmail.com,"1199
Marshall Street ",Whittier,(655) 379-4362
27,Victorina,Radcliff,40,Bryce.Sanders@rogers.ca,"2663
Sutton Street ",Bloomington,(255) 365-0521
```

4. After pressing **Send**, we get the result, as shown in *Figure 13.2*:

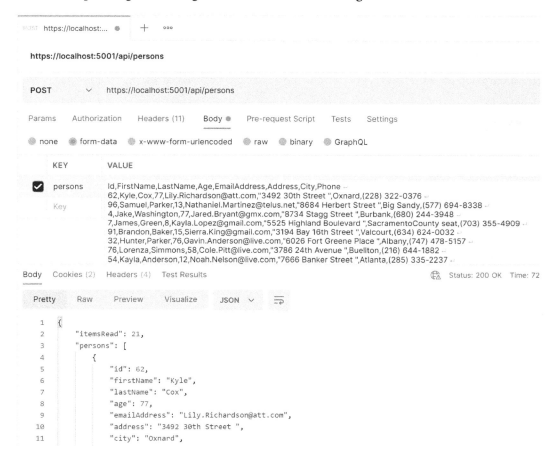

Figure 13.2 – A screenshot of CSV data in Postman

Now, the clients will be able to send CSV-based data to the server.

Summary

This is a good way to transform the input in a way that the action needs. You could also use `ModelBinder` to do some custom validation against the database or whatever you need to do before the model gets passed to the action.

In the next chapter, we will see what you can do with `ActionFilter`.

Further reading

To learn more about `ModelBinder`, you should have a look at the following reasonably detailed documentation:

- Steve Gordon, *Custom ModelBinding in ASP.NET MVC Core*: `https://www.stevejgordon.co.uk/html-encode-string-aspnet-core-model-binding/`

- *Model Binding in ASP.NET Core*: `https://docs.microsoft.com/en-us/aspnet/core/mvc/models/model-binding`

- *Custom Model Binding in ASP.NET Core*: `https://docs.microsoft.com/en-us/aspnet/core/mvc/advanced/custom-model-binding`

14

Creating a Custom ActionFilter

We will keep on customizing on the controller level in this chapter. We'll have a look into action filters and how to create your own `ActionFilter` class to keep your Actions small and readable.

This chapter will cover the following topics:

- Introducing `ActionFilter`
- Using `ActionFilter`

The topics of this chapter belong to the **Model-View-Controller** (**MVC**) layer of the ASP. NET Core architecture, depicted here:

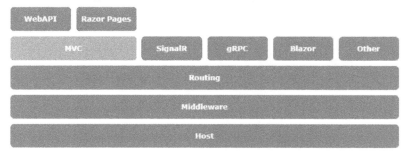

Figure 14.1 – ASP.NET Core architecture

Technical requirements

To follow the exercises in this chapter, you will need to create an ASP.NET Core MVC application. Open your console, shell, or Bash terminal and change to your working directory. Use the following command to create a new MVC application:

```
dotnet new web -n ActionFilterSample -o ActionFilterSample
```

Now, open the project in Visual Studio by double-clicking the project file or in **Visual Studio Code** (**VS Code**) by typing the following command in the already open console:

```
cd ActionFilterSample
code .
```

All of the code samples in this chapter can be found in the GitHub repository for this book at: https://github.com/PacktPublishing/Customizing-ASP. NET-Core-6.0-Second-Edition/tree/main/Chapter14.

Introducing ActionFilter

Action filters are a little bit like middleware because they can manipulate the input and the output but are executed immediately on a specific action or on all actions of a specific controller on the MVC layer, and MiddleWare works directly on the request object on the hosting layer. An ActionFilter class is created to execute code right before or after an action is executed. They are introduced to execute aspects that are not part of the actual action logic: **authorization** is one example of these aspects. AuthorizeAttribute is used to allow users or groups to access specific Actions or Controllers. AuthorizeAttribute is an ActionFilter. It checks whether the logged-on user is authorized or not. If not, it redirects to the login page.

> **Note**
>
> If you apply an ActionFilter globally, it executes on all actions in your application.

The next code sample shows the skeletons of a normal action filter and an async
`ActionFilter`:

```csharp
using Microsoft.AspNetCore.Mvc.Filters;

namespace ActionFilterSample;

public class SampleActionFilter : IActionFilter
{
    public void OnActionExecuting(
        ActionExecutingContext context)
    {
        // do something before the action executes
    }

    public void OnActionExecuted(
        ActionExecutedContext context)
    {
        // do something after the action executes
    }
}

public class SampleAsyncActionFilter : IAsyncActionFilter
{
    public async Task OnActionExecutionAsync(
        ActionExecutingContext context,
        ActionExecutionDelegate next)
    {
        // do something before the action executes
        var resultContext = await next();

        // do something after the action executes;
        // resultContext.Result will be set
    }
}
```

As you can see here, there are always two methods to place code to execute before and after the target action is executed. These action filters cannot be used as attributes. If you want to use action filters as attributes in your Controllers, you need to derive them from `Attribute` or from `ActionFilterAttribute`, as shown in the following code snippet:

```
using Microsoft.AspNetCore.Mvc;
using Microsoft.AspNetCore.Mvc.Filters;


namespace ActionFilterSample;


public class ValidateModelAttribute : ActionFilterAttribute
{
    public override void OnActionExecuting(
        ActionExecutingContext context)
    {
        if (!context.ModelState.IsValid)
        {
            context.Result = new BadRequestObjectResult(
                context.ModelState);
        }
    }
}
```

The preceding code snippet shows a simple `ActionFilter` that always returns `BadRequestObjectResult` if `ModelState` is not valid. This may be useful within a web **application programming interface (API)** as a default check on POST, PUT, and PATCH requests. This could be extended with a lot more validation logic. We'll see how to use it later on.

Another possible use case for an `ActionFilter` is logging. You don't need to log in controller actions directly. You can do this in an action filter to keep your actions readable with relevant code, as illustrated in the following snippet:

```
using Microsoft.AspNetCore.Mvc.Filters;


namespace ActionFilterSample;


public class LoggingActionFilter : IActionFilter
```

```
{
    ILogger _logger;
    public LoggingActionFilter(ILoggerFactory
      loggerFactory)
    {
        _logger =
        loggerFactory.CreateLogger<LoggingActionFilter>();
    }

    public void OnActionExecuting(
        ActionExecutingContext context)
    {
        _logger.LogInformation(
          $"Action '{context.ActionDescriptor.DisplayName}'
            executing");
    }

    public void OnActionExecuted(
        ActionExecutedContext context)
    {
        _logger.LogInformation(
          $"Action '{context.ActionDescriptor.DisplayName}'
            executed");
    }
}
```

This logs an informational message out to the console. You can get more information about the current action out of ActionExecutingContext or ActionExecutedContext—for example, the arguments, the argument values, and so on. This makes action filters pretty useful.

Let's see how action filters work in practice.

Using ActionFilter

Action filters that are actually attributes can be registered as an attribute of an Action or a Controller, as illustrated in the following code snippet:

```
[HttpPost]
[ValidateModel] // ActionFilter as attribute
public ActionResult<Person> Post([FromBody] Person model)
{
    // save the person

    return model; //just to test the action
}
```

Here, we use `ValidateModel` attribute that checks the `ModelState` and returns `BadRequestObjectResult` in case the `ModelState` is invalid; we don't need to check the `ModelState` in the actual Action.

To register action filters globally, you need to extend the MVC registration in the `ConfigureServices` method of the `Startup.cs` file, as follows:

```
builder.Services.AddControllersWithViews()
    .AddMvcOptions(options =>
    {
        options.Filters.Add(new SampleActionFilter());
        options.Filters.Add(new SampleAsyncActionFilter());
    });
```

Action filters registered like this will be executed on every action. This way, you are able to use action filters that don't derive from an attribute.

The `LoggingActionFilter` we created previously is a little more special. It is dependent on an instance of `ILoggerFactory`, which needs to be passed into the constructor. This won't work well as an attribute, because `Attributes` don't support **constructor injection (CI)** via **dependency injection (DI)**. `ILoggerFactory` is registered in the ASP.NET Core DI container and needs to be injected into `LoggingActionFilter`.

Because of this, there are some more ways to register action filters. Globally, we are able to register them as a type that gets instantiated by the DI container, and the dependencies can be solved by the container, as illustrated in the following code snippet:

```
builder.Services.AddControllersWithViews()
    .AddMvcOptions(options =>
    {
        options.Filters.Add<LoggingActionFilter>();
    })
```

This works well. We now have `ILoggerFactory` in the filter.

To support automatic resolution in `Attributes`, you need to use the `ServiceFilter` attribute on the Controller or Action level, as illustrated here:

```
[ServiceFilter(typeof(LoggingActionFilter))]
public class HomeController : Controller
{
```

In addition to the global filter registration, `ActionFilter` needs to be registered in `ServiceCollection` before we can use it with `ServiceFilter` attribute, as follows:

```
builder.Services.AddSingleton<LoggingActionFilter>();
```

To be complete, there is another way to use action filters that needs arguments passed into the constructor. You can use the `TypeFilter` attribute to automatically instantiate the filter. But using this attribute, the filter *isn't* instantiated by the DI container; the arguments need to be specified as arguments of the `TypeFilter` attribute.

See the next snippet from the official documentation:

```
[TypeFilter(typeof(AddHeaderAttribute),
    Arguments = new object[] { "Author", "Juergen Gutsch
        (@sharpcms)" })]
public IActionResult Hi(string name)
{
    return Content($"Hi {name}");
}
```

The type of the filter and the arguments are specified with the `TypeFilter` attribute.

Summary

Action filters give us an easy way to keep actions clean. If we find repeating tasks inside our Actions that are not really relevant to the actual responsibility of the Action, we can move those tasks out to an `ActionFilter`, or maybe a `ModelBinder` or some MiddleWare, depending on how it needs to work globally. The more relevant it is to an Action, the more appropriate it is to use an `ActionFilter`.

There are more kinds of filters, all of which work in a similar fashion. To learn more about the different kinds of filters, reading the documentation is definitely recommended.

In the next chapter, we speed up your web application by using caches.

Further reading

- Microsoft ASP.NET Core filters: `https://docs.microsoft.com/en-us/aspnet/core/mvc/controllers/filters`

15
Working with Caches

In this chapter we will have a look into cache techniques. ASP.NET Core provides multiple ways to cache and we will learn to use and to customize them.

In this chapter, we will be covering the following topics:

- The need for caching
- HTTP-based caching
- Caching using ResponseCachingMiddleware
- Predefining caching strategies using cache profiles
- Caching specific areas using CacheTagHelper
- Caching Manually

The topics in this chapter refer to the MVC layer of the ASP.NET Core architecture:

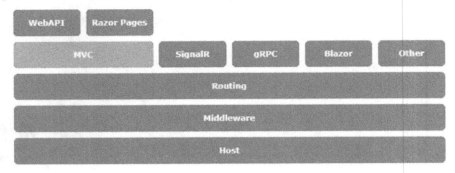

Figure 15.1 – ASP.NET Core architecture

Technical requirements

To follow the descriptions in this chapter, you will need to create an ASP.NET Core MVC application. Open your console, shell, or Bash terminal, and change to your working directory. Use the following command to create a new MVC application:

```
dotnet new mvc -n CacheSample -o CacheSample
```

Now, open the project in Visual Studio by double-clicking the project file or, in VS Code, by typing the following command in the already open console:

```
cd CacheSample
code .
```

All of the code samples in this chapter can be found in the GitHub repository for this book at https://github.com/PacktPublishing/Customizing-ASP. NET-Core-6.0-Second-Edition/tree/main/Chapter15.

Why do we need caching?

Caching speeds up performance, by storing the results in memory or in a distributed cache like a fast Redis database, you can also store cached data in files if it makes sense.

A distributed cache is needed in case you run multiple instances of an application to scale for availability of your application. The instances will run on multiple Docker containers, in a Kubernetes cluster or just on more than one Azure App Services. In that case, the instances should share a cache.

Most application caches are in-memory caches that store data for a short period of time. This is good for most scenarios.

Also, browser do cache the websites or the web applications output. The browsers usually store the entire result in files. As an ASP.NET developer you can control the browsers cache by adding HTTP headers that specify whether the browser should cache or not and that specify how long the cached item should be valid.

A browser cache reduces the number of requests to the server. A cache handling in your code might reduce the number of database access or reduce the access to another time-consuming action.

Both client-side caches and server side caches are useful to increase the performance of your application. Let's have a detailed look at the client side cache.

HTTP-based caching

To control the browsers cache you can set a `Cache-Control` HTTP header. Usually, the `StaticFileMiddleware` doesn't set a Cache-Control header. This means the clients are free to cache how they prefer. If you like to limit the cache time to just one day, you can do this by passing `StaticFileOptions` to the middleware:

```
const string cacheMaxAge = "86400";
app.UseStaticFiles(new StaticFileOptions()
{
    OnPrepareResponse = ctx =>
    {
        ctx.Context.Response.Headers.TryAdd(
            "Cache-Control",
            $"public, max-age={cacheMaxAge}");
    }
});
```

This sets the `Cache-Control` header to every static file that is requested before it gets sent to the client. The `Cache-Control` is set to public, which means it can be publicly cached on every client. The maximum age of the cache items should not be older than 86,400 seconds, which is one day.

Setting the headers to the static file is just an example. You can set the header to every response that needs cache control. You can also disable the cache by setting the Cache-Control header to no-cache.

To learn more about the Cache-Control header, see the following URL: https://datatracker.ietf.org/doc/html/rfc7234#section-5.2

Also, the Expires header might be useful, to specify when the responded content get invalid and should get renewed from the server. See https://datatracker.ietf.org/doc/html/rfc7234#section-5.3

The Vary header specifies a criteria that tells the clients about cache variations. It checks for specific headers to be available. See https://datatracker.ietf.org/doc/html/rfc7231#section-7.1.4

This controls the clients directly via the response object.

Caching using ResponseCachingMiddleware

The ResponseCachingMiddleware caches the responses on the server side and creates responses based on the cached responses. The middleware respects the Cache-Control header in the same way as clients do. That means you are able to control the middleware by setting the specific headers as described in the previews section.

To get it working you need to add the ResponseCachingMiddleware to the Dependency Injection container:

```
builder.Services.AddResponseCaching();
```

And you should use that middleware to the pipeline after the static files and routing got added:

```
app.UseResponseCaching();
```

If you added a CORS configuration, the UseCors method should be called before, as well.

The ResponseCachingMiddleware gets affected by specific HTTP headers. For example, if the Authentication header is set the response doesn't get cached, same with the Set-Cookie header. It also only caches responses that result in a 200 OK result. Error pages and other status codes don't get cached.

You can find the full list of criteria at this URL: https://docs.microsoft.com/en-us/aspnet/core/performance/caching/middleware?view=aspnetcore-6.0#http-headers-used-by-response-caching-middleware.

Using the `ResponseCacheAttribute` on controller level, actions level or pages level you can set the right headers to control the `ResponseCachingMiddleware` by using `ResponseCacheAttribute`:

```
[ResponseCache(Duration = 86400)]
public IActionResult Index()
{
    return View();
}
```

This snippet sets the `Cache-Control` to public with max-age to one day, like the sample in the previews section.

This attribute is pretty powerful, you can also set Vary headers in different ways, as well as the indicator to not cache the output at all. Even a `CacheProfileName` can be set. We are going to have a look at cache profiles in the next section.

These are properties you can set:

- `Duration`: Time range in seconds
- `Location`: The location where to store the cache: Client, Any, or none
- `NoStore`: Disables the cache if it is set to true
- `VaryByHeader`: A header value that varies the cache
- `VaryByQueryKeys`: An array of query key names that varies the cache

Predefining caching strategies using cache profiles

You can predefine caching strategies in a so-called cache profile to reuse them wherever you need. The `CacheProfile` type has the same properties as the `ResponseCache` attribute. To define cache profiles, you need to set options to the MVC services.

In `Program.cs`, the `AddControllersWithViews` method has an overload to configure the `MvcOptions`. Here, you can also add cache profiles:

```
builder.Services.AddControllersWithViews(options =>
{
    options.CacheProfiles.TryAdd("Duration30",
        new CacheProfile
        {
            Duration = 30,
```

```
            VaryByHeader = "User-Agent",
            Location = ResponseCacheLocation.Client
        });
    options.CacheProfiles.TryAdd("Duration60",
        new CacheProfile
        {
            Duration = 60,
            VaryByHeader = "User-Agent",
            Location = ResponseCacheLocation.Client
        });
});
```

You might need to add a `using` statement to `Microsoft.AspNetCore.Mvc`.

This snippet adds two different cache profiles, the first one with a 30 second cache and the second one with a 60 second cache. Both profiles tell the cache to vary by the "User-Agent" header.

To use a profile, you can use the profile name in the response caching attribute:

```
[ResponseCache(CacheProfileName = "Duration30")]
public IActionResult Index()
{
```

Instead of setting all the properties of `ResponseCacheAttribute`, you can just set `CacheProfileName`. Let's see how to use caches the declarative way.

Caching specific areas using CacheTagHelper

You can also cache specific areas of the view. In a scenario where you are not able to cache an entire view, you would be able to just cache a specific area by surrounding it with the `CacheTagHelper`.

To test that, add the following snippet to the `index.cshtml`, that you can find in the `/Views/Home/` folder:

```
<div>
    <p>
        The current time is: @DateTime.Now.ToLongTimeString()
    </p>
</div>
```

```
<cache expires-sliding="@TimeSpan.FromSeconds(7)">
<div>
    <p>
        The current time is: @DateTime.Now.ToLongTimeString()
    </p>
</div>
</cache>
```

This snippet contains two identical p-tags that write out the current time.

The second one is wrapped in a `CacheTagHelper` that has a sliding expiration of 7 seconds defined.

Start the application and see what happens. Navigate to the `Index` page and refresh the browser several times. You will see that only the first time will change while refreshing the page. The second one is cached and stays the same for 7 seconds.

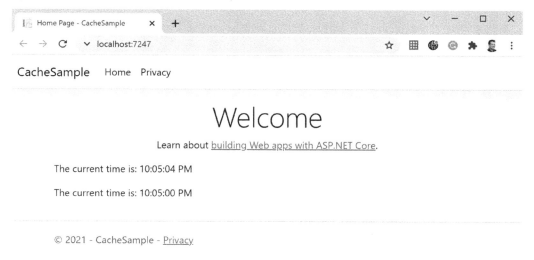

Figure 15.2 - Cached and uncached values

Let's look at what we should do if we need to cache more specifically

Caching Manually

Sometimes it makes sense to cache specifically inside the C# code. For example, if you need to fetch data from an external source or database, it would save time and traffic if you cache the results and don't access the result every time.

Let's try it out by using two different ways to use create and access cache items:

1. To try it out we will extend the `HomeController` a little bit. Start by injecting an instance of the `IMemoryCache` to the controller and store it in a field:

```csharp
using Microsoft.Extensions.Caching.Memory;
public class HomeController : Controller
{
    private readonly ILogger<HomeController> _logger;
    private readonly IMemoryCache _cache;

    public HomeController(
        ILogger<HomeController> logger,
        IMemoryCache cache
        )
    {
        _logger = logger;
        _cache = cache;
    }
```

2. In the `Models` folder, create a file called `Person.cs` and place the following lines in it:

```csharp
namespace CacheSample.Models;

internal class Person
{
    public int Id { get; set; }
    public string? Firstname { get; set; }
    public string? Lastname { get; set; }
    public string? Address { get; set; }
    public string? City { get; set; }
}
```

3. Now we need to add two super complex methods that do some magic for us. Actually, these methods just create fake data and aren't really complex:

```csharp
private IEnumerable<Person>
  LoadDataFromExternalSource()
{
```

```
        return A.ListOf<Person>(10);
    }


    private IDictionary<int, string>
      LoadSuperComplexCalculatedData()
    {
        return Enumerable.Range(0, 10)
            .ToDictionary(
                x => x,
                x => $"Item{Random.Shared.Next()}");
    }
```

The first method uses GenFu that is also used in previous chapters to create a list of Person and fill them with random but valid data. The second method creates a Dictionary of 10 items that also contains random data. The random data make sense to show that the data are actually cached. If the data don't change on the user interface, the data came out of the cache.

4. Type the following command in the project folder to install GenFu:

```
dotnet add package GenFu
```

5. Add the following lines at the beginning of the index action to store the data of the first method in the cache or to load the data out of the cache:

```
if (!_cache.TryGetValue<IEnumerable<Person>>(
    "ExternalSource", out var externalPersons))
{
    externalPersons = LoadDataFromExternalSource();
    _cache.Set(
        "ExternalSource",
        externalPersons,
        new MemoryCacheEntryOptions
        {
            AbsoluteExpiration =
                DateTime.Now.AddSeconds(30)
        });
}
```

This will at first try to load the data out of the cache by using the
ExternalSource cache key. If the cached data doesn't exist, you need to load
them from the original source and store them in the cache using the Set method.

The other way to create and load cached data is to use the GetOrCreate method:

```
var calculatedValues = _cache.GetOrCreate(
    "ComplexCalculate", entry =>
{
    entry.AbsoluteExpiration = DateTime.Now.AddSeconds(30);
    return LoadSuperComplexCalculatedData();
});
```

It works the same way but is pretty much simpler to use. The value that needs to be cached
will be returned in the lambda expression directly while the lambda retrieves the cache
entry that can be configured.

Once the data are there you can return them to the view:

```
return View(new IndexViewModel
        {
            Persons = externalPersons,
            Data = calculatedValues
        });
```

The model that gets returned looks like this:

```
internal class IndexViewModel
{
    public IEnumerable<Person>? Persons { get; set; }
    public IDictionary<int, string>? Data { get; set; }
}
```

Add the next snippet to Index.cshtml right after CacheTagHelper to visualize the
data:

```
<div class="row">
    <div class="col-md-6">
        <ul>
            @foreach (var person in Model.Persons)
            {
```

```
            <li>
                [@person.Id] @person.Firstname @person.Lastname
            </li>
            }
        </ul>
    </div>
    <div class="col-md-6">
        <ul>
            @foreach (var data in Model.Data)
            {
                <li>[@data.Key] @data.Value</li>
            }
        </ul>
    </div>
</div>
```

This creates two lists in two side-by-side columns. Now run the application, call it in the browser, and try to refresh the page. The displayed data shouldn't change even though the data are completely random. Without the cache, the data would change on every reload:

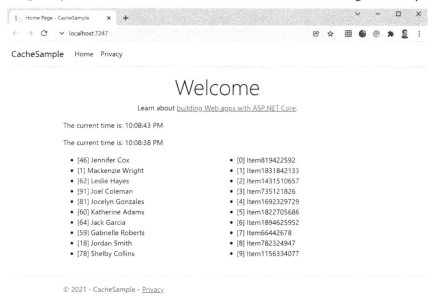

Figure 5.3 - Changing data

That's it. The cache expires every 30 seconds as configured.

Summary

Caches help us to create high performance applications by reducing the calls to resources that are less performant, such as databases, external APIs, or complex calculations. In this chapter, you learned to use the response cache using the `ResponseCachingMiddleware` and the `ResponseCacheAttribute`, and the in-memory cache by using the `CacheTagHelper` as well as by using the `IMemoryCache` manually in the C# code.

In the next chapter, you will learn how to create custom `TagHelper`.

Further reading

More about caching in the ASP.NET Core docs: `https://docs.microsoft.com/en-us/aspnet/core/performance/caching/response?view=aspnetcore-6.0`.

16
Creating Custom TagHelper

In this chapter, we're going to talk about **Tag Helpers**. The built-in `TagHelper` are pretty useful and make Razor much prettier and more readable. Creating custom `TagHelper` will make your life much easier.

In this chapter, we will be covering the following topics:

- Introducing `TagHelper`
- Creating custom `TagHelper`

The topics in this chapter refer to the MVC layer of the ASP.NET Core architecture:

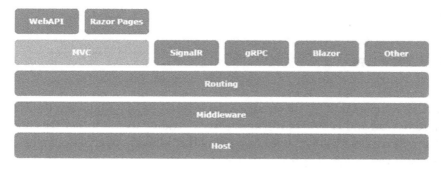

Figure 16.1 – ASP.NET Core architecture

Technical requirements

To follow the examples in this chapter, you will need to create an ASP.NET Core MVC application. Open your console, shell, or Bash terminal and change to your working directory. Use the following command to create a new MVC application:

```
dotnet new mvc -n TagHelperSample -o TagHelperSample
```

Now, open the project in Visual Studio by double-clicking the project file, or in Visual Studio Code by typing the following command in the already-open console:

```
cd TagHelperSample
code .
```

All of the code samples in this chapter can be found in the GitHub repository for this book at: `https://github.com/PacktPublishing/Customizing-ASP.NET-Core-6.0-Second-Edition/tree/main/Chapter16`.

Introducing TagHelper

With **Tag Helpers**, you are able to extend existing HTML tags or create new tags that get rendered on the server side. The extensions or new tags are not visible in browsers. `TagHelper` are a kind of shortcut to write easier (and less) HTML or Razor code on the server side. `TagHelper` will be interpreted on the server and produce "real" HTML code for browsers.

`TagHelper` are not a new thing in ASP.NET Core. They have been present since the framework's first version. Most existing and built-in `TagHelper` are a replacement for the old-fashioned HTML helpers, which still exist and work in ASP.NET Core to keep the Razor views compatible with ASP.NET Core.

A very basic example of extending HTML tags is the built-in `AnchorTagHelper`:

```
<!-- old fashioned HtmlHelper -->
@Html.ActionLink("Home", "Index", "Home")
<!-- new TagHelper -->
<a asp-controller="Home" asp-action="Index">Home</a>
```

Many HTML developers find it a bit strange to have `HtmlHelper` between the HTML tags. It is hard to read and is kind of disruptive while reading the code. Perhaps not for ASP.NET Core developers who are used to reading that kind of code, but compared to `TagHelper`, it is really ugly. `TagHelper` feel more natural and more like HTML, even if they are not, and even if they are getting rendered on the server.

Many HTML helpers can be replaced with a `TagHelper`.

There are also some new tags that have been built with `TagHelper`, tags that are not in HTML but look like HTML. One example is `EnvironmentTagHelper`:

```
<environment include="Development">
    <link rel="stylesheet"
        href="~/lib/bootstrap/dist/css/bootstrap.css" />
    <link rel="stylesheet" href="~/css/site.css" />
</environment>
<environment exclude="Development">
    <link rel="stylesheet"
        href="https://ajax.aspnetcdn.com/ajax/bootstrap/
            3.3.7/css/bootstrap.min.css"
        asp-fallback-href=
            "~/lib/bootstrap/dist/css/bootstrap.min.css"
        asp-fallback-test-class="sr-only"
        asp-fallback-test-property="position"
        asp-fallback-test-value="absolute" />
    <link rel="stylesheet"
        href="~/css/site.min.css"
        asp-append-version="true" />
</environment>
```

This `TagHelper` renders (or doesn't render) the contents depending on the current runtime environment. In this case, the target environment is the development mode. The first environment tag renders the contents if the current runtime environment is set to `Development`, and the second one renders the contents if it is *not* set to `Development`. This makes it a useful helper in rendering debuggable scripts or styles in `Development` mode and minified and optimized code in any other runtime environment.

Let's now see how we can create our own custom `TagHelper`.

Creating custom Tag Helpers

To use all the custom `TagHelper` that we will create in this chapter, you need to refer to the current assembly to tell the framework where to find the `TagHelper`. Open the `_ViewImports.cshtml` file in the `View/` folder and add the following line at the end of the file:

```
@addTagHelper *, TagHelperSample
```

Here's a quick example showing how to extend an existing tag using a `TagHelper`:

1. Let's assume we need to have a tag configured in bold and colored in a specific color:

    ```
    <p strong color="red">Use this area to provide
      additional information.</p>
    ```

    This looks like pretty old-fashioned HTML from the 90s, but this is just to demonstrate a simple `TagHelper`.

2. The current method to do this task is to use a `TagHelper` to extend any tag that has an attribute called `strong`, as shown in the following code snippet:

    ```
    using Microsoft.AspNetCore.Razor.TagHelpers;

    namespace TagHelperSample.TagHelpers;

    [HtmlTargetElement(Attributes = "strong")]
    public class StrongTagHelper : TagHelper
    {
        public string Color { get; set; }

        public override void Process(
            TagHelperContext context,
            TagHelperOutput output)
        {
            output.Attributes.RemoveAll("strong");

            output.Attributes.Add("style",
                "font-weight:bold;");
    ```

```
        if (!String.IsNullOrWhiteSpace(Color))
        {
            output.Attributes.RemoveAll("style");
            output.Attributes.Add("style",
                $"font-weight:bold;color:{Color};");
        }
    }
}
```

The first line tells the tag helper to work on tags with a target attribute of `strong`. This `TagHelper` doesn't define its own tag, but it does provide an additional attribute to specify the color.

The `Process` method defines how to render the HTML to the output stream. In this case, it adds some CSS inline styles to the current tag. It also removes the target attribute from the current tag. The `color` attribute won't show up.

This will appear as follows:

```
<p style="font-weight:bold;color:red;">Use this area
    to provide additional information.</p>
```

The next example shows how to define a custom tag using a `TagHelper`:

1. Let's create this class, called `GreeterTagHelper`:

    ```
    using Microsoft.AspNetCore.Razor.TagHelpers;

    namespace TagHelperSample.TagHelpers;

    public class GreeterTagHelper : TagHelper
    {
        [HtmlAttributeName("name")]
        public string Name { get; set; }

        public override void Process(
            TagHelperContext context,
            TagHelperOutput output)
    ```

```
    {
        output.TagName = "p";
        output.Content.SetContent($"Hello {Name}");
    }
2
```

2. This `TagHelper` handles a `greeter` tag that has a property name. In the `Process` method, the current tag will be changed to a `p` tag and the new content is set as the current output:

```
<greeter name="Readers"></greeter>
```

The result looks like this:

```
<p>Hello Readers</p>
```

But what if you need to do something a bit more complicated? Let's explore further.

Examining a more complex scenario

The `TagHelper` in the last section were pretty basic, simply designed to show how `TagHelper` work. The next example is a little more complex and shows a real scenario. This `TagHelper` renders a table with a list of items. This is a generic `TagHelper` and shows a real reason to create your own custom `TagHelper`. With this, you are able to reuse an isolated piece of view code. For example, you can wrap **Bootstrap** components to make them much easier to use with just one tag, instead of nesting five levels of `div` tags. Alternatively, you can just simplify your Razor views:

1. Let's start by creating the `DataGridTagHelper` class. This next code snippet isn't complete, but we will complete the `DataGridTagHelper` class in the following steps:

```
using Microsoft.AspNetCore.Razor.TagHelpers;

namespace TagHelperSample.TagHelpers;

public class DataGridTagHelper : TagHelper
{
    [HtmlAttributeName("Items")]
    public IEnumerable<object> Items { get; set; }

    public override void Process(
```

```
        TagHelperContext context,
        TagHelperOutput output)
    {
        output.TagName = "table";
        output.Attributes.Add("class", "table");
        var props = GetItemProperties();

        TableHeader(output, props);
        TableBody(output, props);
    }
}
```

In the `Process` method, we call private sub-methods that do the actual work to make the class a little more readable.

You might need to add the following `using` statements at the beginning of the file:

```
using System.Reflection;
using System.ComponentModel;
```

2. Because this is a generic `TagHelper`, incoming objects need to be analyzed. The `GetItemProperties` method gets the type of the property items and loads the `PropertyInfo` from the type. `PropertyInfo` will be used to get the table headers and the values:

```
private PropertyInfo[] GetItemProperties()
{
    var listType = Items.GetType();
    Type itemType;
    if (listType.IsGenericType)
    {
        itemType = listType.GetGenericArguments()
            .First();
        return itemType.GetProperties(
            BindingFlags.Public |
            BindingFlags.Instance);
    }
    return new PropertyInfo[] { };
}
```

3. The following code snippet shows the generation of the table headers.
    The `TableHeader` method writes the requisite HTML tags directly to
    `TagHelperOutput`. It also uses the list of `PropertyInfo` to get the property
    names that will be used as table header names:

```
private void TableHeader(
    TagHelperOutput output,
    PropertyInfo[] props)
{
    output.Content.AppendHtml("<thead>");
    output.Content.AppendHtml("<tr>");
    foreach (var prop in props)
    {
        var name = GetPropertyName(prop);
        output.Content.AppendHtml($"<th>{name}</th>");
    }
    output.Content.AppendHtml("</tr>");
    output.Content.AppendHtml("</thead>");
}
```

4. Using property names as table header names is not always useful. This is
    why the `GetPropertyName` method also tries to read the value from
    `DisplayNameAttribute`, which is part of the `DataAnnotation` that is heavily
    used in data models that are displayed in MVC user interfaces. Therefore, it makes
    sense to support this attribute:

```
private string GetPropertyName(
    PropertyInfo property)
{
    var attribute = property
        .GetCustomAttribute<DisplayNameAttribute>();
    if (attribute != null)
    {
        return attribute.DisplayName;
    }
    return property.Name;
}
```

5. Also, values need to be displayed. The `TableBody` method does that job:

```
private void TableBody(
    TagHelperOutput output,
    PropertyInfo[] props)
{
    output.Content.AppendHtml("<tbody>");
    foreach (var item in Items)
    {
        output.Content.AppendHtml("<tr>");
        foreach (var prop in props)
        {
            var value = GetPropertyValue(prop, item);
            output.Content.AppendHtml(
                $"<td>{value}</td>");
        }
        output.Content.AppendHtml("</tr>");
    }
    output.Content.AppendHtml("</tbody>");
}
```

6. To get the values from the actual object, the `GetPropertyValue` method is used:

```
private object GetPropertyValue(
    PropertyInfo property,
    object instance)
{
    return property.GetValue(instance);
}
```

7. To use this `TagHelper`, you just need to assign a list of items to this tag:

```
<data-grid items="Model.Persons"></data-grid>
```

In this case, it is a list of people, which we get in the `Persons` property of our current model.

8. The `Person` class we are using here looks like this:

```
using System.ComponentModel;
```

```
namespace TagHelperSample.Models;


public class Person
{
    [DisplayName("First name")]
    public string FirstName { get; set; }

    [DisplayName("Last name")]
    public string LastName { get; set; }

    public int Age { get; set; }

    [DisplayName("Email address")]
    public string EmailAddress { get; set; }
}
```

Not all of the properties have `DisplayNameAttribute`, so the fallback in the `GetPropertyName` method is needed to get the actual property name instead of the `DisplayName` value.

Put the `Person` class into a `Person.cs` inside the `Models` folder.

9. You also need a service to load the data into the `Index` action of the `HomeController`. Create a `Services` folder and place a file called `PersonService.cs` into it. Put the following snippet inside the file:

```
using TagHelperSample.Models;
using GenFu;


namespace TagHelperSample.Services;


public interface IService
{
    IEnumerable<Person> AllPersons();
}
internal class PersonService : IService
{
    public IEnumerable<Person> AllPersons()
    {
```

```
        return A.ListOf<Person>(25);
    }
}
```

Here, again, we use GenFu to auto-generate the list of persons. If you didn't already install it, you need to execute the following command to load the NuGet package:

```
dotnet add package GenFu
```

If this is done you should add PersonService to ServiceCollection in the Program.cs file:

```
builder.Services.AddTransient<IService, PersonService>();
```

And, last but not least, PersonService should be used in HomeController:

```
using Microsoft.AspNetCore.Mvc;
using TagHelperSample.Models;
using TagHelperSample.Services;

namespace TagHelperSample.Controllers;

public class HomeController : Controller
{
    private readonly IService _service;

    public HomeController(
        IService service)
    {
        _service = service;
    }

    public IActionResult Index()
    {
        ViewData["Message"] = "Your application
            description page.";

        var persons = _service.AllPersons();
        return View(new IndexViewModel
        {
            Persons = persons
```

```
        });
    }
```

10. This `TagHelper` needs some more checks and validations before you can use it in production, but it works. It displays a list of fake data that is generated using `GenFu` (see *Chapter 12, Content Negotiation Using a Custom OutputFormatter*, to learn about `GenFu`):

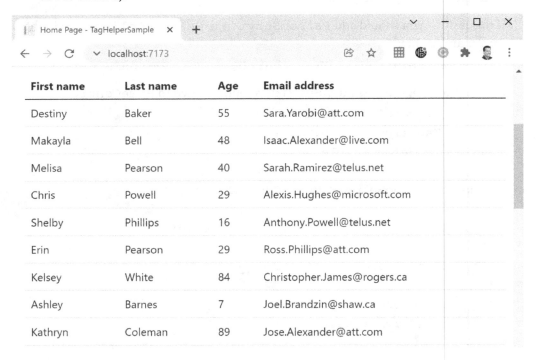

Figure 16.2 – The TagHelper sample in action

Now, you are able to extend this `TagHelper` with a lot more features, including sorting, filtering, and paging. Feel free to try it out in a variety of contexts.

Summary

Tag Helpers are pretty useful when it comes to reusing parts of the view and simplifying and cleaning up your views, as in the example with `DataGridTagHelper`. You can also provide a library with useful view elements. There are some more examples of pre-existing `TagHelper` libraries and samples that you can try out in the *Further reading* section.

This is the last chapter of the second edition of *Customizing ASP.NET Core*. We're glad you read all the chapters. We hope you found the chapters useful and that they will help you optimize your applications.

Further reading

- Damian Edwards, *TagHelperPack*: `https://github.com/DamianEdwards/TagHelperPack`

- David Paquette, *TagHelperSamples*: `https://github.com/dpaquette/TagHelperSamples`

- *TagHelpers for Bootstrap by Teleric*: `https://www.red-gate.com/simple-talk/dotnet/asp-net/asp-net-core-tag-helpers-bootstrap/`

- *TagHelpers for jQuery*: `https://www.jqwidgets.com/asp.net-core-mvc-tag-helpers/`

Index

U

W

X

Packt.com

Subscribe to our online digital library for full access to over 7,000 books and videos, as well as industry leading tools to help you plan your personal development and advance your career. For more information, please visit our website.

Why subscribe?

- Spend less time learning and more time coding with practical eBooks and Videos from over 4,000 industry professionals

- Improve your learning with Skill Plans built especially for you

- Get a free eBook or video every month

- Fully searchable for easy access to vital information

- Copy and paste, print, and bookmark content

Did you know that Packt offers eBook versions of every book published, with PDF and ePub files available? You can upgrade to the eBook version at packt.com and as a print book customer, you are entitled to a discount on the eBook copy. Get in touch with us at customercare@packtpub.com for more details.

At www.packt.com, you can also read a collection of free technical articles, sign up for a range of free newsletters, and receive exclusive discounts and offers on Packt books and eBooks.

Other Books You May Enjoy

If you enjoyed this book, you may be interested in these other books by Packt:

ASP.NET Core and Vue.js

Devlin Basilan Duldulao

ISBN: 978-1-80323-279-9

- Discover CQRS and mediator pattern in the ASP.NET Core 5 Web API
- Use Serilog, MediatR, FluentValidation, and Redis in ASP.NET
- Explore common Vue.js packages such as Vuelidate, Vuetify, and Vuex
- Manage complex app states using the Vuex state management library
- Write integration tests in ASP.NET Core using xUnit and FluentAssertions
- Deploy your app to Microsoft Azure using the new GitHub Actions for continuous integration and continuous deployment (CI/CD)

C# 10 and .NET 6 – Modern Cross-Platform Development – Sixth Edition

Mark J. Price

ISBN: 978-1-80107-736-1

- Build rich web experiences using Blazor, Razor Pages, the Model-View-Controller (MVC) pattern, and other features of ASP.NET Core

- Build your own types with object-oriented programming

- Write, test, and debug functions

- Query and manipulate data using LINQ

- Integrate and update databases in your apps using Entity Framework Core, Microsoft SQL Server, and SQLite

- Build and consume powerful services using the latest technologies, including gRPC and GraphQL

- Build cross-platform apps using .NET MAUI and XAML

Packt is searching for authors like you

If you're interested in becoming an author for Packt, please visit `authors.packtpub.com` and apply today. We have worked with thousands of developers and tech professionals, just like you, to help them share their insight with the global tech community. You can make a general application, apply for a specific hot topic that we are recruiting an author for, or submit your own idea.

Share Your Thoughts

Now you've finished *Customizing ASP.NET Core 6.0*, we'd love to hear your thoughts! Scan the QR code below to go straight to the Amazon review page for this book and share your feedback or leave a review on the site that you purchased it from.

`https://packt.link/r/1803233605`

Your review is important to us and the tech community and will help us make sure we're delivering excellent quality content.

www.ingramcontent.com/pod-product-compliance
Lightning Source LLC
Chambersburg PA
CBHW060600060326
40690CB00017B/3774